ISEKI-Food Series

Series editor: Kristberg Kristbergsson *University of Iceland, Reykjavík, Iceland*

More information about this series at http://www.springer.com/series/7288

Anna McElhatton • Mustapha Missbah El Idrissi
Editors

Modernization of Traditional Food Processes and Products

 Springer

Editors
Anna McElhatton
University of Malta
Msida, MSD2090, Malta

Mustapha Missbah El Idrissi
Université Mohamed V
École Normale Supérieure
Rabat, Morocco

Series Editor
Kristberg Kristbergsson
University of Iceland
Reykjavík, Iceland

Integrating Food Science and Engineering Knowledge Into the Food Chain
ISBN 978-1-4939-7947-9   ISBN 978-1-4899-7671-0   (eBook)
DOI 10.1007/978-1-4899-7671-0

Springer New York Heidelberg Dordrecht London
© Springer Science+Business Media New York 2016
Softcover re-print of the Hardcover 1st edition 2016

Printed on acid-free paper

Springer Science+Business Media LLC New York is part of Springer Science+Business Media
(www.springer.com)

Series Preface

The ISEKI-Food Series was originally planned to consist of six volumes of texts suitable for food science students and professionals interested in food safety and environmental issues related to sustainability of the food chain and the well-being of the consumer. As the work progressed, it soon became apparent that the interest and need for texts of this type exceeded the topics covered by the first six volumes published by Springer in 2006–2009. The series originate in work conducted by the European thematic network "ISEKI-Food," an acronym for "Integrating Safety and Environmental Knowledge In to Food Studies." Participants in the ISEKI-Food network come from most countries in Europe, and most of the institutes and universities involved with food science education at the university level are represented. The network was expanded in 2008 with the ISEKI Mundus program with 37 partners from 23 countries outside of Europe joining the consortium, and it continues to grow with approximately 200 partner institutions from 60 countries from all over the world in 2011. Some international companies and nonteaching institutions have also participated in the program. The network was funded by the ERASMUS program of the EU from 1998 to 2014 first as FoodNet coordinated by Professor Elisabeth Dumoulin at AgroParisTech-site de MASSY in France. The net then became known as "ISEKI-Food" and was coordinated by Professor Cristina Silva at the Catholic University of Portugal, College of Biotechnology (Escola) in Porto, from 2002 to 2011 when Professor Paola Pittia at the University of Teramo in Italy became coordinator of ISEKI-Food 4.

The main objectives of ISEKI-Food have been to improve the harmonization of studies in food science and engineering in Europe and to develop and adapt food science curricula emphasizing the inclusion of safety and environmental topics. The program has been further expanded into the ISEKI-Food Association (https://www.iseki-food.net/), an independent organization devoted to the objectives of the ISEKI consortium to further the safety and sustainability of the food chain through education. The motto of the association is "Integrating Food Science and Engineering Knowledge into the Food Chain." The association will continue work on the ISEKI-Food Series with several new volumes to be published in the near future. The series was continued with volume 7 in 2012 with the publication of *Novel Technologies in*

Food Science: Their Impact on Products, Consumer Trends and The Environment, edited by Anna McElhatton and Paolo J. Sobral. The book is intended for food scientists and engineers and readers interested in the new and emerging food processing technologies that are intended to provide foods that are safe, but maintain most of their original freshness. All 13 chapters are written from a safety and environmental standpoint with respect to the emerging technologies.

We now see the publication of the *Trilogy of Traditional Foods* written for food science professionals as well as for the interested general public. The trilogy is in line with the internationalization of the ISEKI consortium and will offer close to 80 chapters dedicated to different traditional foods from all over the world. The trilogy starts with a text offering general descriptions of different traditional foods and topics related to consumers and sensory aspects with a volume entitled *Traditional Foods: General and Consumer Aspects* edited by the undersigned and Jorge Oliveira. The second book in the trilogy is *Modernization of Traditional Food Processes and Products* edited by Anna McElhatton and Mustapha Missbah El Idrissi. The chapters are devoted to recent changes and modernizations of specific traditional foods focusing on the processing and engineering aspects. The third volume in the trilogy is *Functional Properties of Traditional Foods* devoted to functional and biochemical aspects of traditional foods and the beneficial effects of bioactive components found in some traditional foods.

The series will continue with several books including the textbook *Food Processing* edited by the undersigned and Semih Ötles, intended for senior-level undergraduates and junior graduate students providing a comprehensive introduction to food processing. The book should also be useful to professionals and scientists interested in food processing both from the equipment and process approach as well as in the physicochemical aspect of food processing. The book will contain five sections starting with chapters on the basic principles and physicochemical properties of foods, followed by sections with chapters on conversion operations, preservation operations, and food processing operations with separate chapters on most common food commodities. The final section will be devoted to post-processing operations.

Applied Statistics for Food and Biotechnology, edited by Gerhard Schleining, Peter Ho, and Saverio Mannino, will be intended for graduate students and industry personnel who need a guide for setting up experiments so that the results will be statistically valid. The book will provide numerous samples and case studies on how to use statistics in food and biotechnology research and testing. It will contain chapters on data collection, data analysis and presentation, handling of multivariate data, statistical process control, and experimental design.

Process Energy in Food Production, edited by Winfried Russ, Barbara Sturm, and the undersigned, will offer an introduction section on basic thermodynamics and an overview of energy as a global element and environmental effects of energy provision and usage. This will be followed with chapters on the use of energy in various food processes like flour production, bakery, fish processing, meat processing, brewery and beverage production, direct and indirect heat integration in breweries, fruit juice, spray drying systems (milk powder), and chilling and storage of fresh horticultural products. There will also be chapters related to energy supply (thermal, solar, hydroelectric), energy distribution, insulation for energy saving, storage systems for heat and coldness, waste heat recovery, and energy management systems.

Three more volumes are being prepared. The textbook *Physical Chemistry for Food Scientists* edited by Stephan Drusch and Kirsi Jouppila; the book will provide a text for the senior undergraduate- and beginning graduate-level students on the basic principles of physical chemistry of foods. The first part of the book will be devoted to fundamental principles of physical chemistry. The second part of the book will be devoted to the physical chemistry of food systems. A volume entitled *Consumer-Driven Development of Food for Health and Well Being* edited by the undersigned, Paola Pittia, Margarida Vieira, and Howard R. Moskowitz is in preparation with chapters on general aspects of food development, the house of quality and Stage-Gate® process, consumer aspects of food development, mind genomics, conceptualization of well-being in the framework of food consumption, formulation of foods in the development of food for health and well-being, ingredients contribution for health and well-being, new trends on the extension of shelf life, nutritional aspects of development of foods focusing on health and well-being, regulatory and policy aspects, and several case studies on product development with special emphasis on health and well-being. Finally there is a *Book on Ethics in Food Production and Science* that will be edited by Rui Costa and Paola Pittia being developed.

The ISEKI-Food Series draws on expertise from universities and research institutions all over the world, and we sincerely hope that it my offer interesting topics to students, researchers, professionals, as well as the general public.

Reykjavík, Iceland   Kristberg Kristbergsson
July, 2011

Series Preface to Volumes 1–6

The single most important task of food scientists and the food industry as a whole is to ensure the safety of foods supplied to consumers. Recent trends in global food production, distribution, and preparation call for increased emphasis on hygienic practices at all levels and for increased research in food safety in order to ensure a safer global food supply. The ISEKI-Food book series is a collection of books where various aspects of food safety and environmental issues are introduced and reviewed by scientists specializing in the field. In all of the books, a special emphasis was placed on including case studies applicable to each specific topic. The books are intended for graduate students and senior-level undergraduate students as well as professionals and researchers interested in food safety and environmental issues applicable to food safety.

The idea and planning of the books originates from two working groups in the European thematic network "ISEKI-Food," an acronym for "Integrating Safety and Environmental Knowledge In to Food Studies." Participants in the ISEKI-Food network come from 29 countries in Europe, and most of the institutes and universities involved with food science education at the university level are represented. Some international companies and nonteaching institutions have also participated in the program. The ISEKI-Food network is coordinated by Professor Cristina Silva at the Catholic University of Portugal, College of Biotechnology (Escola) in Porto. The program has a web site http://www.esb.ucp.pt/iseki/. The main objectives of ISEKI-Food have been to improve the harmonization of studies in food science and engineering in Europe and to develop and adapt food science curricula emphasizing the inclusion of safety and environmental topics. The ISEKI-Food network started on October 1 in 2002 and has recently been approved for funding by the EU for renewal as ISEKI-Food 2 for another 3 years. ISEKI has its roots in an EU-funded network formed in 1998 called FoodNet where the emphasis was on casting a light on the different food science programs available at the various universities and technical institutions throughout Europe. The work of the ISEKI-Food network was organized into five different working groups with specific task all aiming to fulfill the main objectives of the network.

The first four volumes in the ISEKI-Food book series come from WG2 coordinated by Gerhard Schleining at Boku University in Austria and the undersigned. The main task of the WG2 was to develop and collect materials and methods for teaching of safety and environmental topics in the food science and engineering curricula. The first volume is devoted to food safety in general with a practical and a case study approach. The book is composed of 14 chapters which were organized into three sections on preservation and protection, benefits and risk of microorganisms, and process safety. All of these issues have received high public interest in recent years and will continue to be in the focus of consumers and regulatory personnel for years to come. The second volume in the series is devoted to the control of air pollution and treatment of odors in the food industry. The book is divided into eight chapters devoted to defining the problem, recent advances in analysis, and methods for prevention and treatment of odors. The topic should be of special interest to industry personnel and researchers due to recent and upcoming regulations by the European Union on air pollution from food processes. Other countries will likely follow suit with more strict regulations on the level of odors permitted to enter the environment from food processing operations. The third volume in the series is devoted to utilization and treatment of waste in the food industry. Emphasis is placed on sustainability of food sources and how waste can be turned into by products rather than pollution or landfills. The book is composed of 15 chapters starting off with an introduction of problems related to the treatment of waste and an introduction to the ISO 14001 standard used for improving and maintaining environmental management systems. The book then continues to describe the treatment and utilization of both liquid and solid wastes with case studies from many different food processes. The last book from WG2 is on predictive modeling and risk assessment in food products and processes. Mathematical modeling of heat and mass transfer as well as reaction kinetics is introduced. This is followed by a discussion of the stoichiometry of migration in food packaging, as well as the fate of antibiotics and environmental pollutants in the food chain using mathematical modeling and case study samples for clarification.

Volumes 5 and 6 come from work in WG5 coordinated by Margarida Vieira at the University of Algarve in Portugal and Roland Verhé at Gent University in Belgium. The main objective of the group was to collect and develop materials for teaching food safety-related topics at the laboratory and pilot plant level using practical experimentation. Volume 5 is a practical guide to experiments in unit operations and processing of foods. It is composed of 20 concise chapters each describing different food processing experiments outlining theory, equipment, procedures, applicable calculations, and questions for the students or trainee followed by references. The book is intended to be a practical guide for the teaching of food processing and engineering principles. The final volume in the ISEKI-Food book series is a collection of case studies in food safety and environmental health. It is intended to be a reference for introducing case studies into traditional lecture-based safety courses as well as being a basis for problem-based learning. The book consists of 13 chapters containing case studies that may be used, individually or in a series, to discuss a range of food safety issues. For convenience the book was divided into three main sections with the first devoted to case studies, in a more general framework with a number of specific issues in safety and health ranging from acrylamide and nitrates to botulism and listeriosis. The second section is devoted to some well-known outbreaks related to food intake in different countries. The final section of the book takes on food safety from the perspective of the researcher. Cases are based around experimental data and examine the importance of experimental planning, design, and analysis.

The ISEKI-Food books series draws on expertise from close to 100 universities and research institutions all over Europe. It is the hope of the authors, editors, coordinators, and participants in the ISEKI network that the books will be useful to students and colleagues to further their understanding of food safety and environmental issues.

Reykjavík, Iceland   Kristberg Kristbergsson
March, 2008

Preface

Modernization of Traditional Food Processes and Products is the second book in the *Trilogy of Traditional Foods*, part of the ISEKI-Food *Series*. The three books in the trilogy are devoted to different characteristics of traditional foods. The trilogy covers general and consumer aspects, modernization of traditional foods, and functional properties of traditional foods in a total of 74 chapters written by authors from all over the world. The "List of Contributors" in this second book of the trilogy cites authors from China, Thailand, India, Argentina, New Zealand, and the United Kingdom and thus embraces the ethos of the ISEKI-Food family.

Modernization of Traditional Food Processes and Products is divided into two sections, with the first section dedicated to foodstuffs developed in Europe. The first six chapters in this section deal with foods and beverages that contain dairy products, while Chaps. 7 and 8 deal with grain-based products. The second section is devoted to foods from the rest of the world. Products covered in these chapters are very varied, as are the processes that produce them.

The book should be of interest to the practicing food professional and the interested reader, with all chapters written by food scientists or engineers working in industry or involved in academia. The book is meant to encourage both food professionals and the interested general public to look beyond the label and packaging of the foodstuffs tackled in the 15 chapters. By providing insight into the history and development of these foods and beverages, the editors hope to encourage people to become even more curious about the food products they commonly use, especially the composition and origins of these items.

Msida, Malta Anna McElhatton
Rabat, Morocco Mustapha Missbah El Idrissi

Acknowledgments

ISEKI_Food 3 and ISEKI_Mundus 2 were thematic networks on food studies, funded by the European Union through the Lifelong Learning and Erasmus Mundus programs as projects No. 142822-LLP-1-2008-PT-ERASMUS-ENW and 145585-PT-2008-ERA MUNDUS—EM4EATN, respectively. The ISEKI_Mundus 2 project was established to contribute for the internationalization and enhancement of the quality of the European higher education in food studies and work toward the network sustainability, by extending the developments undergoing through the Erasmus Academic Network ISEKI Food 3 to other countries and developing new activities toward the promotion of good communication and understanding between European countries and the rest of the world.

Education and Culture DG

Lifelong Learning Programme

Education and Culture DG

ERASMUS MUNDUS

Education and Training

Contents

Contributors

L.R. Alieva FSAEI HPE, North-Caucasus Federal University, Stavropol, Russia

Lorena Atarés Instituto de Ingeniería de Alimentos para el Desarrollo, Universidad Politécnica de Valencia, Valencia, Spain

John D. Brooks Faculty of Health and Environmental Sciences, School of Applied Sciences, Auckland University of Technology, Auckland, New Zealand

Karl Georg Busch Beuth University of Applied Sciences, Berlin, Germany

M.C. Bustos Instituto de Ciencia y Tecnología de los Alimentos Córdoba (CONICET—Universidad Nacional de Córdoba), Córdoba, Argentina

Amparo Chiralt Instituto de Ingeniería de Alimentos para el Desarrollo, Universidad Politécnica de Valencia, Valencia, Spain

József Csanádi Faculty of Engineering, University of Szeged, Szeged, Hungary

S. Drusch Institute of Food Technology and Food Chemistry, Technische Universität Berlin, Berlin, Germany

U. Einhorn-Stoll Institute of Food Technology and Food Chemistry, Technische Universität Berlin, Berlin, Germany

Dedi Fardiaz Department of Food Science and Technology, Bogor Agricultural University—IPB, Bogor, Indonesia

Wunwiboon Garnjanagoonchorn Department of Food Science and Technology, Faculty of Agro-Industry, Kasetsart University, Bangkok, Thailand

Jiashun Gong Faculty of Food Science and Technology, Yunnan Agricultural University, Kunming, P.R. China

Gudmundur Gudmundsson Department of Food Science and Nutrition, University of Iceland, Reykjavik, Iceland

LYSI hf., Fiskislod 5-9, Reykjavik, Iceland

Cecilia Hodúr Faculty of Engineering, University of Szeged, Szeged, Hungary

Kristberg Kristbergsson Department of Food Science and Nutrition, University of Iceland, Reykjavik, Iceland

Leonardus Broto Sugeng Kardono Research Centre for Chemistry, Indonesian Institute of Sciences—LIPI, Serpong, Indonesia

I.K. Kulikova FSAEI HPE, North-Caucasus Federal University, Stavropol, Russia

Zsuzsanna László Faculty of Engineering, University of Szeged, Szeged, Hungary

Andrzej Lenart Faculty of Food Sciences, Warsaw University of Life Sciences— Szkoła Główna Gospodarstwa Wiejskiego (WULS-SGGW), Warsaw, Poland

A.E. León Instituto de Ciencia y Tecnología de los Alimentos Córdoba (CONICET—Universidad, Nacional de Córdoba), Córdoba, Argentina

Qi Lin Faculty of Food Science and Technology, Yunnan Agricultural University, Kunming, P. R. China

Michelle Lucke-Hutton Faculty of Health and Environmental Sciences, School of Applied Sciences, Auckland University of Technology, Auckland, New Zealand

C.S. Martínez Instituto de Ciencia y Tecnología de los Alimentos Córdoba (CONICET—Universidad Nacional de Córdoba), Córdoba, Argentina

Anna McElhatton Faculty of Health Sciences, University of Malta, Msida, Malta

Peter C. Mitchell Northern Ireland Centre for Food and Health, University of Ulster, Coleraine, Northern Ireland, UK

P.S. Minz National Dairy Research Institute, Karnal, India

O.I. Oleshkevich FSAEI HPE, North-Caucasus Federal University, Stavropol, Russia

Hadi K. Purwadaria Department of Food Technology, Swiss German University— SGU, Serpong, Indonesia

Nick Roskruge Faculty of Health and Environmental Sciences, School of Applied Sciences, Auckland University of Technology, Auckland, New Zealand

R.R.B. Singh National Dairy Research Institute, Karnal, India

Sarote Sirisansaneeyakul Department of Biotechnology, Faculty of Agro-Industry, Kasetsart University, Chatuchak, Bangkok, Thailand

S.E. Vinogradskaya FSAEI HPE, North-Caucasus Federal University, Stavropol, Russia

Qiuping Wang Faculty of Food Science and Technology, Yunnan Agricultural University, Kunming, P.R. China

Department of Biotechnology, Faculty of Agro-Industry, Kasetsart University, Bangkok, Thailand

Małgorzata Ziarno Faculty of Food Sciences, Warsaw University of Life Sciences — Szkoła Główna Gospodarstwa Wiejskiego (WULS-SGGW), Warsaw, Poland

About the Editors

Anna McElhatton is a senior lecturer and head of the Department of Food Studies and Environmental Health in the Faculty of Health Sciences at the University of Malta. Anna McElhatton has an undergraduate degree in pharmacy, an M.A. in bioethics from the University of Malta, and M.Phil. and Ph.D. from the Queen's University of Belfast in Northern Ireland. Her main research interests include food safety (of dairy products), modernization of traditional foods, sensory aspects of food preference, and ethical issues in research.

Mustapha Missbah El Idrissi is professor of general microbiology at the "Ecole Normale Supérieure" at Mohammed V University of Rabat (UM5R), Morocco. He is responsible of the master's degree "biotechnology and environment." He was a senior lecturer of food microbiology for over 20 years at Mohamed Premier University in Oujda and moved to Rabat to teach general microbiology at UM5R. His main research include legumes improvement through biological nitrogen fixation with rhizobia. The molecular characterization of bacteria in different ecosystems is also one of his research interests.

About the Series Editor

Kristberg Kristbergsson is Professor of Food Science at the Department of Food Science and Nutrition at the University of Iceland. Dr. Kristbergsson has a B.S. in food science from the Department of Chemistry at the University of Iceland and earned his M.S., M.Phil., and Ph.D. in food science from Rutgers University. His research interests include physicochemical properties of foods, new processing methods for seafoods, modernization of traditional food processes, safety and environmental aspects of food processing, extrusion cooking, shelf-life and packaging, biopolymers in foods, and delivery systems for bioactive compounds.

Part I
Europe

Chapter 1
Traditional Polish Curd Cheeses

Małgorzata Ziarno and Andrzej Lenart

Contents

M. Ziarno (✉) • A. Lenart
Faculty of Food Sciences, Warsaw University of Life Sciences—Szkoła Główna Gospodarstwa Wiejskiego (WULS-SGGW), Nowoursynowska Str. 159C, 02-776 Warsaw, Poland
e-mail: malgorzata_ziarno@sggw.pl

© Springer Science+Business Media New York 2016
A. McElhatton, M.M. El Idrissi (eds.), *Modernization of Traditional Food Processes and Products*, Integrating Food Science and Engineering Knowledge Into the Food Chain 11, DOI 10.1007/978-1-4899-7671-0_1

1.1   Introduction

Curd cheese is produced throughout the world, usually by denaturation and coagulation of milk proteins according to traditional recipes and rules. This is a very
large and diverse group of dairy products. Together with the ripening cheeses, they
are one of the most varied and attractive direction of milk processing. Products of
this type in Poland are commonly referred to as "white cheese," and it is hard to
find their counterpart in the world, although somewhat similar to them are
American "pressed farmers cheese" and German "quark." Polish curd cheeses are
characterized by a full, delicate, creamy taste, and a creamy, dense consistency
with smooth grains.

Curd cheeses have a significant traditional niche in Poland and remain a popular
foodstuff. Homemade curd cheeses were in the past produced by most home cooks
in village households. They had a unique aroma and a place in Polish cuisine and
culture. In 2010, the average consumption of curd cheese in Poland was 6.6 kg per
capita and was higher than in other countries of the European Union, and their production exceeded ca. 370,000 t. They retain the reputation of being a natural product
of high nutritional value that makes these products attractive and able to survive
very well in a competitive dairy market. This could be attributed to the nutritional
attributes of the curd cheese (i.e., a high content of protein compounds, easily
absorbable calcium, B vitamins, in particular riboflavin, and easily digestible fat),
combined with an uncomplicated production process, the affordable price of the
final product, and the wide application of curd cheese in households (Śmietana et al.
1994a, b, c; Kitlas and Ziarno 2002; Ziarno and Zaręba 2007; Siemianowski and
Szpendowski 2014). The unique role of curd cheese is one of the characteristics that
distinguish Polish cuisine from all other.

Curd cheeses are offered in several convenient packages and forms for customers: full-fat, skimmed or semi skimmed, natural, steamed, fried, smoked or flavored,
formed in wedge or cubes, sliced or ground, wrapped in parchment paper or plastic
wrap, in cups or vacuum packed, with or without sour cream added. They are sold
in various portions: 200 g, 250 g, even 1 or 2 kg packed. They are suitable for direct
consumption, and as ingredients in savory dishes or sweets such as cakes. The best-
selling formats are vacuum packed; on the other hand, curd cheese packed only in
parchment paper and sold by weight is no longer appealing to consumers.

Manufacturers of curd cheese have taken care to improve both the quality of
the cheeses as well as the design of convenience packaging. The quality of curd
cheese and their attractiveness to consumers are dependent on many factors. The
most important are the quality of the raw milk used for production, hygienic conditions of the production process, and the conditions of storage and distribution
that guarantee shelf life as these cheeses do not contain preservatives. In addition, the traditional curd cheese aroma and ease to use are important quality attributes that need to be retained if consumers are to continue appreciating this
traditional product.

1.2   Traditional Technology and Modern Modification of Curd Cheese Production

Curd cheese is a term that describes a large and diverse group of dairy products that are produced in large quantities. A common feature of all curd cheese is their processing, which is the coagulation of milk protein (mainly casein) by lactic acid fermentation or an acid–rennet combination (a coagulant enzyme simultaneously in conjunction with the lactic acid bacteria) (Śmietana et al. 1994a, b, c; Żylińska et al. 2014). The main stages of the production of curd are (Guinee et al. 1999; Bohdziewicz 2008):

- Preparation of milk for processing
- Treatment and coagulation of milk
- Cutting and pressing the curd
- Formation of the curd cheese

Detailed curd cheese production technologies differ in (Nitecka and Popiołek 1990; Śmietana et al. 2003):

- Use of different types of coagulation agent (rennet, acid),
- Duration of milk coagulation
- Fragmentation of the clot
- Forming techniques

The resulting final products are intended for direct consumption (Śmietana et al. 1994a, b, c; Pikul 2004; Siemianowski et al. 2013a).

1.2.1   Preparation of Milk for Processing

Milk used to production of curd cheese should be of high quality and have appropriate chemical composition (especially casein) (Bohdziewicz 2008; Danków and Pikul 2011).

The first step is the preparation of milk for processing (traditional homemade curd cheese was produced from raw milk, but curd cheese production in industrial conditions requires the use of pasteurized milk). Skim or fat standard milk is subjected to low pasteurization (Low-Temperature-Long-Time, LTLT) or high (High-Temperature-Short-Time, HTST or Very High Temperature, VHT). In the first case, the milk is heated to a temperature of 65 °C and held for 30 min. HTST requires higher temperatures, a minimum of 72 °C and a holding time of 15 s. In VHT systems, milk is heated to 80–90 °C for 2–25 s (or even 10–15 min) before it is cooled.

LTLT pasteurization has been used in industrial scale production for many decades in production lines that use the batch production method. However, currently HTST or VHT systems are used in continuous flow pasteurization of milk.

The significant difference in these pasteurization processes is the degree of denaturation and coagulation of albumin and globulins of milk, which in the case of the HTST is as high as 50 %. This is important in the production of curd cheese as it improves the structure of the formed curd (Lopez et al. 1995; Śmietana et al. 2003). VHT system can further increase the yield of the curd cheese by 10–20 % due to coagulation of most of the whey protein with casein (Śmietana et al. 1994a, b, c; Siemianowski et al. 2013b).

1.2.2   Treatment and Coagulation of Milk

Curd cheese production is highly dependent on the appropriate mechanical and thermal treatment of milk curd. A curd formed from milk proteins, mainly casein, coagulates at pH that corresponds to the value of isoelectric point of casein (Ziarno and Zaręba 2007; Siemianowski and Szpendowski 2012). Milk preparation and coagulation and curd treatment including cutting and mixing all take place in temperature-controlled baths or tanks.

There are two types of coagulation: long- and short-time. In the long-time coagulation method, pasteurized milk is cooled to 20–23 °C, starter lactic acid bacteria are added (inoculum of 0.5–2.5 % of processed milk), and milk is left in these conditions for 14–16 h to obtain a curd (Śmietana et al. 1994a, b, c; Żylińska et al. 2014). This method requires the use of mesophilic lactic acid bacteria: *Lactococcus lactis* subsp. *lactis*, *Lactococcus lactis* subsp. *cremoris*, and *Lactococcus lactis* subsp. *diacetylactis* (Żylińska et al. 2014).

The presence of aroma-producing bacteria facilitates the development of the traditional Polish curd cheese. *Lactococcus lactis* subsp. *diacetylactis* produces lactic acid and diacetyl in addition to CO_2. Large amounts of gas produced cause that the curd cut floats on the surface of cheese whey and does not sink to the bottom.

In short-time method, pasteurized milk is cooled to 32–35 °C and then starter lactic acid bacteria are added (inoculum of 5 % of processed milk), including thermophilic bacteria such as *Streptococcus thermophilus*. In this case, the coagulation time is only 6–8 h (Guinee et al. 1999). However, the product obtained by this method is less aromatic due to insufficient growth of aroma-producing bacteria (Śmietana et al. 1994a, b, c; Guinee et al. 1999; Śmietana et al. 2003).

The pH value of curd obtained is about 4.5–4.7, and acidity 30–34°SH (sometimes the milk is acidified to only 20–25°SH and coagulation is achieved by heating to 40 °C). Mature curd has the consistency of soft jelly with no cracks and separation of whey. It should give a break with equal edges and smooth walls (Dmytrów et al. 2010).

1.2.3   Cutting, Pressing, and Formation of the Curd

The curd treatment process is designed to release the whey from the milk and to obtain a concentrated protein curd (Nitecka and Popiołek 1990; Siemianowski et al. 2013b).

The cutting of curd should involve a process of gradual size reduction. The curd should be sliced into blocks of about 12×12 cm. The curds should not be crushed. Gentle further size reduction to 1–5 cm is carried out with gentle mixing while heating to 35 °C (temperature increase by about 1 °C per 5 min). This dries the grain curd and facilitates the separation of clear whey (Śmietana et al. 1994a, b, c; Guinee et al. 1999). Depending on the type of curd cheese produced, a separation of all or part of the whey followed by the addition of the same volume of water (heated to 30–35 °C) is recommended; this whole volume of curd and water is then gently mixed.

After cooling the curd grains to 10–12 °C, the curd is transferred to cone-shaped flax bags, molds covered with a flax cheese cloth, or in a perforated plastic molds; this facilitates further drainage of whey from the curd. The flax bags or molds covered with a flax cheese cloth have been used both in traditional manufacture of curd cheese and for homemade curd cheese. Industrialization of curd cheese production and high standards of hygiene requirements necessitated the use of perforated plastic molds.

Pressing also takes place in flax bags or molds. It begins with the application of gentle pressure, not exceeding 10 N/kg, and then the pressure is increased up to about 30 N/kg. Pressing the curd is a process that causes significant further drying of the curd (Śmietana et al. 1994a, b, c; Siemianowski et al. 2013b).

Curd cheese can be formed or molded in the form of a flat block or cylinder, or as flat wedge shape ("klinek" in Polish language). The curd cheese has to be chilled to 2–8 °C soon after packing to prevent over acidification of curd.

1.3   Polish Classification of Curd Cheese

The Polish classification system describes the following types of unripened curd cheese (Ziarno and Zaręba 2007; Bohdziewicz 2008):

- Acid curd cheese (cut into cube or wedge shapes, full-fat, half-fat, or skimmed).
- Curd cheese or acid–rennet cheese (produced by traditional method with mild treatment of a curd), with or without sour cream added, homogenized.
- Acid–rennet curd cheese (produced by centrifugal method), with uniform structure, natural, or flavored.
- Acid–rennet curd cheese similar to American cottage cheese.
- Acid curd cheese produced from milk heated to 92 °C for few seconds, with addition of 0.04 % $CaCl_2$, with traditional curd treatment.
- Milk–buttermilk and buttermilk curd cheese (obtained from a mixture of milk and buttermilk or only buttermilk) with traditional method of curd treatment.

There are also ripened curd cheeses that are produced from skimmed curd cheese through the crushing of the curd, salting, and seasoning with cumin or pepper, followed by maturation for up to twelve days. This process of ripening is called "gliwienie" in Polish and refers to the process of growth of proteolytic bacteria and mold of *Oospora lactis* species on the surface of curd cheese (Bohdziewicz 2008).

1.4   Curd Cheese Properties

The final product has a taste and a smell that is described as mild, clean, slightly acidic. Its structure and texture are uniform, compact, without lumps, and slightly loose, and it may be slightly granular. The color of curd cheese should be white to light cream and be uniform throughout the whole cheese.

Water content in curd cheese should not exceed 70–75 %. Acidity should be 80–110°SH.

Traditional curd cheeses have a short shelf life, usually less than 72 h, due to the presence of live lactic acid bacteria and high water content (70–78 %) (Cais and Wojciechowski 1996). The application of modern technology and modern methods of curd cheese packaging allows extending the shelf life to 21 days (Śmietana et al. 2003; Olborska and Lewicki 2006; Siemiankowski et al. 2012).

1.5   Curd Cheese on the List of Polish Traditional Food Products

The National List of Traditional Products has been established by virtue of an act in 2004. One of the main aims of this list was the promotion of Polish traditional products made with the use of traditional methods. This list is published and maintained by the Minister of Agriculture and Rural Development as well as governors of particular Polish provinces. It is some guide through the Polish regional cuisine and the source of information about Polish regional traditions, ways of food production, and the characteristics of the products listed there. A dozen products from that list are protected within the EU scheme as PDOs, PGIs, and TSGs. In the middle of 2015, The National List of Traditional Products included 1433 food products, including 84 traditional Polish dairy products from all Polish regions (The National List of Traditional Products 2015). The most important traditional curd cheeses are presented below.

1.5.1   Ser Zabłocki

It was registered on the List of Traditional Food on 5th March 2009 in the category of dairy products. Zabłocki cheese is a traditional product that has been produced for over a hundred years in the Southern Podlasie region, where the monks were involved in the production of cheese. The cheese is obtained only from fresh cow's milk (The National List of Traditional Products 2015).

Zabłocki cheese is produced as a small curd cheese in shape of cylinder (length approx. 20 cm, weight 75 g, width 15 cm, height 4 cm). Taste: Slightly salty. The consistency was slightly spongy, plastic, and allows users to easily cut with a knife. Color is white, slightly creamy with a slight sheen. Smell is buttery (The National List of Traditional Products 2015).

1.5.2   Gzik Wielkopolski

It was registered on the List of Traditional Food on 11th May 2007 in the category of dairy products. In the eighteenth and nineteenth centuries, it contributed to the increase in the manufacture of cheese from cow's milk in Wielkopolska region, where curd cheese called "gzik" was used to season dishes (The National List of Traditional Products 2015). There are many recipes for gzik; one of the easiest is to blend the curd cheese with sour cream, chives or onions, and salt. This mixture is then served with the potatoes.

Gzik wielkopolski is a white curd cheese that has a thick and sticky consistency (The National List of Traditional Products 2015).

1.5.3   Łódzki Twaróg Tradycyjny

It was registered on the List of Traditional Food on 15th February 2010 in the category of dairy products. Twaróg tradycyjny is produced from cow's milk and formed into cubes and wedges; it is either white or light cream in color that is uniform throughout the cheese mass. It is easily cut with a knife. The cube is usually about 10 cm long, 8 cm wide, and about 4 cm high with a weight of 250–380 g or about 17 cm long, 8 cm high, and about 7 cm wide with a weight of 0, 5–0, 8 kg. The weight of wedge is 150–500 g. The consistency is uniform, compact, without lumps, and slightly loose to easily cut with a knife. Taste and smell are clean, soft, and slightly acidic with a pasteurized flavor. Color should be white to light cream, and uniform throughout the mass (The National List of Traditional Products 2015).

1.5.4   Ser Biały Z Handzlówki

It was registered on the List of Traditional Food on 19th August 2010 in the category of dairy products. Ser biały z Handzlówki is a traditional product produced in the Podkarpacie region. It is a white or cream colored curd cheese, with a shape of hook or circular which is sometimes cut into quarters of approximately 0.5–0.6 kg. Its consistency is lumpy or creamy, depending on the fat content. Taste and smell are sweet–sour, and the aroma is characteristic of an acid curd (The National List of Traditional Products 2015).

1.5.5   Serek Twarogowy Ziarnisty

It was registered on the List of Traditional Food on 17th January 2011 in the category of dairy products. It is a traditional product produced in the Podkarpacie region. The cheese consists of cheese grains mixed with sour cream. The shape of cheese grains is irregular. Cheese size is 0.2–20 kg. Taste and smell are creamy; it may be slightly sour and slightly salty. Color is white or light cream, uniform throughout the mass (The National List of Traditional Products 2015).

1.5.6   Twaróg Hajnowski

It was registered on the List of Traditional Food on 14th July 2009 in the category of dairy products. It is a traditional product produced in the North-Eastern part of Podlasie region—a part of the Bialowieza National Park, an area with unique microclimate that guarantees the character of product. This area is a typical agricultural area that has a strong tradition of dairy farming. Therefore, twaróg hajnowski is produced in a way which is common in other parts of Poland, but it owes its specific quality to the milk quality of the area. Twaróg hajnowski is produced in the form of block or wedge, of weight 600–800 g and 250 g, respectively. Its consistency is uniform, compact, without lumps, and slightly loose. The taste is clean, mild, and slightly sour. Color is white to light cream, depending on fat content (The National List of Traditional Products 2015).

1.5.7   Jędrzejowski Twarożek Śmietankowy

It was registered on the List of Traditional Food on 26th July 2010 in the category of dairy products. It is a traditional product produced in the Świętokrzyskie region. The origin of this curd cheese is closely connected with the Regional Dairy Cooperative, which was founded in February 1937. This is the kind of acid–rennet curd cheese. Initially, it was produced using wood equipment from pasteurized milk, then in standard dairy equipment after the modernization of the cooperative in 1957. It has a shape of cube of weight 30–80 g. Its consistency is thick, creamy with a smooth grain—lubricating. Taste and smell are slightly acidic, aromatic, and delicate. Color is white to light cream, uniform throughout the mass (The National List of Traditional Products 2015).

1.5.8   Wielkopolski Twaróg Tradycyjny

It was registered on the List of Traditional Food on 1st December 2010 in the category of dairy products. It is a traditional product produced in the Wielkopolska region, a typical agricultural area that is free of heavy industry. Wielkopolski twaróg tradycyjny is produced from cow's milk and is the simplest form of cheese produced, with a method that has remained unchanged to this day. Freshly collected milk is poured into bowls, jars, or pots and left at room temperature for spontaneous fermentation. The fermentation occurs at 20–30 °C due to the presence of the environmental flora. During the fermentation cream accumulates on the surface of the curd, and depending on how fatty cheese should be, sour cream may be collected, then the curd is processed by heating. With heating the whey separates. The curd is then poured into flax cloth bags or kerchiefs to remove the whey and form cheese shape. The appearance of this curd cheese is uniform, compact, with granular mass. Its shape is rectangular cut by hand, wedge, or oval-molded by hand. The net weight of the block is not less than 100 g. The consistency is grainy. Taste and smell are slightly acidic, clean, and mild. The color is white to light cream, depending on fat content (The National List of Traditional Products 2015).

1.5.9   Serek Śmietankowy Wielkopolski

It was registered on the List of Traditional Food on 12th August 2010 in the category of dairy products. It is a traditional product produced in the Wielkopolska region where the first Polish dairy cooperative was established in 1882. This is acid–rennet curd cheese with chives or onion. Milk is heated to 30 °C and then lactic acid bacteria culture is added, and after about 4 h rennet is added. The milk is left to form curds (approximately 12 h). The resulting curd is cut and inverted several times to remove the whey. The final product is a white color curd cheese with irregular shape, resembling a cylinder, slightly ellipsoidal, of weight 250 g. Its consistency is uniform, compact, and without lumps. Taste and smell are slightly acidic, and gentle (The National List of Traditional Products 2015).

References

Bohdziewicz K (2008) Acid curds—a processing. Przegląd Mleczarski 7:12–15. Abstract in English
Cais D, Wojciechowski J (1996) Changes in selected qualitative characteristics of curd cheeses during their storage. Przegląd Mleczarski 6:177–179. Abstract in English
Danków R, Pikul J (2011) Technological suitability of goat milk for processing. Nauka Przyr Technol 5(2):1–15. Abstract in English
Dmytrów I, Mituniewicz-Małek A, Dmytrów K (2010) Physicochemical and sensory features of acid curd cheese (tvarog) produced from goat's milk and mixture of cow's and goat's milk. Food Sci Technol Qual 2(69):46–61. Abstract in English
Guinee TP, Pudja PD, Farkye NY (1999) Fresh acid-curd cheese varieties. In: Fox PF (ed) Cheese: chemistry, physics and microbiology, vol 2. Chapman & Hall, London, pp 363–419
Kitlas M, Ziarno M (2002) Trial of fortification of quark with calcium salts. Food Sci Technol Qual 3(32):79–88. Abstract in English
Lopez MB, Botet MJ, Hellin P, Luna A, Laencina J (1995) Effect of thermal treatment on goat milk clotting time. Milchwissenschaft 50(3):126–129
Nitecka E, Popiołek P (1990) Effect of coagulation of milk to changes in the nutritional value of the curd cheese protein. Przemysł Spożywczy 44(11):284–286. Abstract in English
Olborska K, Lewicki PP (2006) Packing process for chosen dairy products as a critical control point. Inżynieria Rolnicza 7(82):351–358. Abstract in English
Pikul J (2004) Factors affecting the shelf life of milk and milk products. Part 2. Dairy products with short, intermediate and long shelf life. Chłodnictwo 10(39):41–47. Abstract in English

Siemiankowski K, Szpendowski J, Bohdziewicz K (2012) Effect of packaging, storage and distribution conditions on the extension of shelf life of acid twaróg cheese. Chłodnictwo 47(9):38–41. Abstract in English

Siemianowski K, Szpendowski J (2012) Possibilities of tvarog cheeses enrichment with calcium in the light of hitherto existing research. Eng Sci Technol 4(7):83–98. Abstract in English

Siemianowski K, Szpendowski J (2014) Importance of tvorog in human nutrition. Problemy Higieny i Epidemiologii 95(1):115–119. Abstract in English

Siemianowski K, Szpendowski J, Bohdziewicz K, Kołakowski P, Bardowski J (2013a) Nutritional value of acid tvarog produced from milk concentrated by evaporation and ultrafiltration (UF). Eng Sci Technol 4(11):111–119. Abstract in English

Siemianowski K, Szpendowski J, Bohdziewicz K, Kołakowski P, Pawlikowska K, Żylińska J, Bardowski J (2013b) Effect of the dry matter content in milk on the composition and sensory properties of acid tvarog cheese. Folia Pomeranae Universitatis Technologiae Stetinensis. Agric Aliment Pisc Zootech 302(25):113–124. Abstract in English

Śmietana Z, Szpendowski Z, Bohdziewicz K, Świgoń J (1994a) General production rules of tvorog and curd cheeses. Part I. Przegląd Mleczarski 1:7–9. Abstract in English

Śmietana Z, Szpendowski Z, Bohdziewicz K, Świgoń J (1994b) General production rules of tvorog and curd cheeses. Part II. Przegląd Mleczarski 2:41–43. Abstract in English

Śmietana Z, Szpendowski Z, Bohdziewicz K, Świgoń J (1994c) General production rules of tvorog and curd cheeses. Part III. Przegląd Mleczarski 3:69–71. Abstract in English

Śmietana Z, Szpendowski J, Bohdziewicz K (2003) Characteristics of traditional Polish curd cheese obtained by its own design and technology. Przegląd Mleczarski 4:126–129. Abstract in English

The National List of Traditional Products (2015) The Polish Ministry of Agriculture and Rural Development. http://www.minrol.gov.pl. Update 2015.07.08 /in Polish/

Ziarno M, Zaręba D (2007) Curd cheeses are not so bad as they are described. Higiena 3–4(28):16–17. In Polish

Żylińska J, Siemianowski K, Bohdziewicz K, Pawlikowska K, Kołakowski P, Szpendowski J, Bardowski J (2014) Starter cultures for acid curd—role and expectations. Post Mikrobiol 53(3):288–298. Abstract in English

Chapter 2
"Túró Rudi": Everyone's Favourite Milk Dessert in Hungary

József Csanádi, Zsuzsanna László, and Cecilia Hodúr

Contents

2.1   The History of "Túró Rudi"

The history of Túró Rudi goes back to 1954 when three dairy industry professionals from Hungary—an industrial manager, a food industry engineer and a factory supervisor—took a 2-week-long study trip to the Soviet Union to observe the Soviet styled methodologies of dairy production. It was there that they first saw a product which could be considered a predecessor of today's Túró Rudi. Having no information about the name of this soft, round confection made from a mixture of lactic acid curd, butter and fat (and then sugared and coated in chocolate), they simply called it túró mignon. This was probably where they got the idea to develop a similar product which could appeal to the Hungarian palate. The túró mignon was also mentioned in their report on the study trip, published in the pages of Hungarian Dairy Industry Newsletter.

Practical development of the new product first took place under the direction of the production manager (Rudolf Mandeville) and his small team at the factory (in Budapest). Many people mistakenly attribute the name "Rudi" to Mandeville, but the name actually came from Sándor Klein, a young psychologist enlisted to create

J. Csanádi (✉) • Z. László • C. Hodúr
Faculty of Engineering, University of Szeged, Szeged, Hungary
e-mail: csanadi@mk.u-szeged.hu

© Springer Science+Business Media New York 2016
A. McElhatton, M.M. El Idrissi (eds.), *Modernization of Traditional Food Processes and Products*, Integrating Food Science and Engineering Knowledge Into the Food Chain 11, DOI 10.1007/978-1-4899-7671-0_2

the packaging design, and who was responsible for the initial advertising campaign as well. The original Túró Rudi wrapping, decorated with the image of a red, pigtailed little girl, was developed by two of Klein's students at the College of Applied Arts and displayed the characteristic dotted pattern right from the start. Other anecdotal information implies that Klein, one of the original teams said: "Let's call it Túró Rudi", which and promptly incurred the hatred of newspaper publishers, who considered the brand name obscene and refused to advertise it. In spite of this—or maybe because of it—Túró Rudi became a huge success.

Mass production, as we mentioned, was launched at a Dairy Firm located in Budapest in 1968, but the operation was soon transferred to Szabolcs County due to the poor manufacturing conditions, e.g. lack of space and insufficient environment. The original wooden production equipment was first disassembled and then transported to the city of Nyíregyháza, where it was then reassembled prior to a trial run that lasted for several weeks. Production continued here until the 20th of August 1970, when a new factory was established in the city of Mátészalka, but a separate department for Túró Rudi production only was first set up only during the 1980s.

At first, the product was sold in the vicinity of Mátészalka and also in Budapest, and then it was distributed further in the western regions as production capacity increased. Túró Rudi production in the city of Nagybánhegyes was launched in 1981, which was a source of healthy competition for the Dairy Firm in the city of Mátészalka up to the 1990s, when both plants continued operating under the auspices of the Nutricia Group. During the early 1980s, like in Mátészalka, the products still had a brief shelf life of only 3–4 days, but this increased to 14 days in the course of further development, which also led to the introduction of several new flavours in addition to the basic non-flavoured Túró Rudi produced in the beginning. So, Túró Rudi with authentic flavours (walnut, peanut) and their chocolate-coated versions were first manufactured in Nagybánhegyes, but as product development continued during the 1970s numerous different products were introduced, mainly throughout the 1980s and 1990s.

After 1990s, an ongoing development was evident not only in terms of the product itself but also in packaging, technical improvement and advertising. In the beginning, products in Mátészalka were packaged by hand, much like the way in which traditional Hungarian Christmas sweets were enclosed by twisting both ends of the foil wrapper. Later, this method was replaced by more up-to-date machinery and packaging materials, including the tinfoil wrapping utilized during the 1980s and the consumer-friendly opening strip introduced after 2000. "Pöttyös" products have been regularly displayed at various trade fairs and food product exhibitions over the last few decades, winning nearly twenty different awards recognizing their quality, uniqueness and innovation. Besides these developments and undiminished popularity of Pöttyös Túró Rudi, the production of similar products by other factory owners started to increase the competition.

Since 2000, the brand has undergone continuous rejuvenation, in the course of which not only its packaging but also its message has been given a fresh and youthful image. Along with its domestic success, Pöttyös (Dotted) has also penetrated foreign markets outside Hungary. Túró Rudi was introduced to Romania and Slovakia in 2004 under the name Dots and as of 2006, the brand also became available for customers in the Italian Auchan stores, after which it gained tremendous popularity in Spain as well. In 2007, two new innovations took place with the introduction of the brands Pöttyös Túró Bonbon and Pöttyös Ice-cream, and the latter one is only available on seasonal basis. The owner of the brand Túró Rudi is the "Bonafarm" Group which includes Sole-Mizo Hungary Corporation as producer. Nowadays, the "newest" trend in the Túró Rudi production is the return to the original mix composition, without modern colouring agents and other discussed additives.

2.2 Technology

2.2.1 Traditional Technology

The technology of Túró Rudi is built on the production of túró. Túró technology basically determines the texture, sensory properties and economy of this product. The main operations of túró making are clotting by lactic-acid starters and drainage of curd. First of all, we would like to introduce the traditional túró and Túró Rudi technology.

The traditional túró and Túró Rudi technology used the approach and technologies in use during the 1970s. Critical steps required for food safety and economical production to compensate for variability in raw milk quality were adopted and included HTST pasteurization (High Temperature Short Time). The conditions used were 72–78 °C temperature with 40 sec. holding time. Holding, drainage and cooling of curd were usually slow and the whole process was mostly manual with numerous workers were employed for mixing, forming and packaging the product. Túró Rudi could therefore be called as a handmade product. The technology was divided into three sections as follows:

- Milk pre-treatment
- Túró making
- Mixing, forming and packaging.

During pre-treatment, milk first was filtered, separated to cream and skim milk. Pasteurization of milk and cream in plate heat exchanger followed the separation at 72–78 °C (milk) and 95–105 °C (cream). Then fat standardization was performed by mixing of pasteurized and cooled cream and skim milk, occasionally. Firstly, non-fat túró was usually made and the fat standardization was performed later while mixing and flavouring the drained curd (Szakály 2001).

Then milk temperature was set to 24–26 °C and it was pumped to a traditional, opened cheese vat. It was followed by mixing the starter and the fermentation until pH 4.6 when milk was coagulated by the increased organic acid content. After coagulation, the curd was cut into small pieces with tools equipped with wires. Cutting was terminated as soon as the pieces of curd were of correct size (3–4 cm diameter, called "walnut"-size) and the cutting tools were exchanged to stirring tools. The syneresis (shrinking of curd pieces) immediately began after the cutting of curd which then resulted in the appearance of whey. It was followed by heating up the curd–whey mix to 42–50 °C to accelerate the "drying of curd". The curd was stirred only a few times and gently in favour of temperature balance and to avoid the huge curd loss. While heating, "drying" period usually lasted 3–5 h, not all the bacteria were killed, so the acidity of curd usually increased further (Fenyvessy and Csanádi 2007).

The high total solid content of túró is very important in terms of product texture. Thus, the curd was separated from whey in perforated vats (called "curd-drainage car" since it had wheels). The small-meshed nets in drainage vats prevented the curd from passing through. Curd drainage in the cheese making room or in a better case, in a cooling room, continued while the total solids of curd reached the appropriate value (about 16–24 h). Sometimes, the curd was pressed to remove more whey.

After it a mixture was created in a simple mixer from acid curd (túró), butter and flavouring agents like sugar and lemon oil. This was the original recipe of Túró Rudi but the precise mixing ratio was (is) a close kept secret. The traditional Túró Rudi technology is demonstrated in Fig. 2.1.

Compression and forming were performed in mincer-like equipment, and then the formed mix was cut into small pieces and dipped into melted chocolate or chocolate-like coating material. As a next step, the pieces were collected in trays and cooled in a cooling room (5 °C) and then it was followed by packaging. Many workers were needed in the finishing operations, thus the labour costs were considerable.

2.2.2 Modern Túró Rudi Technology

Modernization of the process used to manufacture Túró Rudi technology has sought to eliminate difficulties associated with the traditional process. Furthermore, there were also attempts to increase product yield…

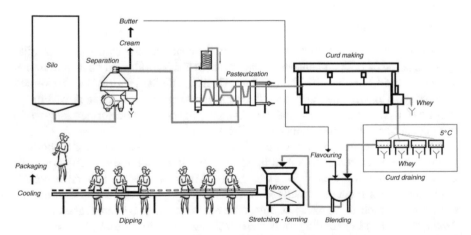

Fig. 2.1 Flow chart of traditional Túró Rudi processing (Modified from Dairy Handbook, Alfa-Laval Engineering)

Deficiencies related to food safety were addressed through milk. Raw milk quality is considered very high in Hungary with more than 95 % of all raw cow milk meeting EU Requirements with the larger producers selecting premium quality raw milk for Túró Rudi production. Recontamination and excessive acidification have to be avoided during production; this has been successfully addressed through the use of closed cheese tanks and rapid curd cooling used in many dairy firms for this purpose (Walstra et al. 2006).

Use of a modern closed cheese vat in curd making and curd handling prevents recontamination of milk or curd with microbes and other contaminants, but the visual monitoring of the stages of curd making cannot be monitored as well as in a traditional open cheese vat, thus not all Túró Rudi producers use it.

In the past two decades (1990–2010), significant research has been carried out to find means of increasing the yield of curd. Use of the HTST heat treatment at high temperature (85–95 °C) with the incorporation of β-lactoglobulin to the curd is a common practice that slightly increases curd yield (or lower protein loss in whey). Use of ultrafiltration (UF) and protein concentration resulted in significant improvements related to the yield (Britz and Robinson 2008). The evaporation of retentate from UF, then lactic acid gelation resulted in higher yields and therefore a more economical production. There were further improvements associated with the texture of the product which became creamier. This creamy inner texture did not gain unanimous success among consumers. (Modern production of Túró Rudi making is demonstrated in Fig. 2.2).

The product yield could also be increased further by selection of milk of higher protein content. New flavoured and jam-filled product versions have also been developed and put on the market and in general demand has remained stable.

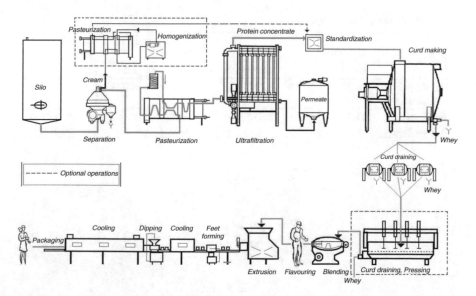

Fig. 2.2 Flow chart of modern Túró Rudi processing (Modified from Dairy Handbook Alfa-Laval Engineering)

Further changes in technology include acceleration of curd drainage to get the required total solids of curd in less time. To obtain low fat or non-fat túró, perforated rotary drums are used for the fast drainage of curd. Pressing of túró is also widespread thus increasing total solids, as the high total solids of túró are very important in standardization of the composition and texture properties of the end product. The separated whey may be further exploited such that its residual protein content is retrieved. This retrieval is achieved through a process of ultrafiltration under conditions that remove sugars, followed by evaporation. This process produces whey protein concentrate powder that can be reintegrated into production thus optimizing protein recovery.

The finishing operations have also changed. Modern cutters-mixers are used for the fast mixture preparation, thus the texture will not be as creamy as in a standard blinder with long blending or when only the ultra-filtered milk concentrate is used for production.

High quality extruders used for extrusion can contain combined forming head for stuffed products, thus the inner structure of product can be controlled by demand. Cutting of the extruded mixture to the appropriate size is processed automatically. Dipping, setting of chocolate-like coating (or chocolate) and cooling are processed by a continuous processing line using circulation system for coating and a cooling tunnel. Products coming out of the tunnel are immediately wrapped by machines and moved into a cooling room. The typical composition of **Túró Rudi** (30 g) is similar to that of a medium fat cheese (Table 2.1) and these days is packaged very attractively (Fig. 2.3) In conclusion, the modern technologies adopted by Túró Rudi production have led to improved food safety levels and the efficient and therefore economical and value-added production of a traditional product brought into the twenty-first century.

Table 2.1 Composition of Túró Rudi

Product name	Nutrition values (g/100 g)			Energy in 100 g product		Energy in one product	
	Protein	Fat	Carbohydrates	kJ	kcal	kJ	kcal
Túró Rudi 30 g (dark coating)	10.79	18.94	38.86	1544.9	369.1	463.5	110.7
Túró Rudi 30 g (milk coating)	11.20	22.32	39.15	1681.9	402.3	504.6	120.7

Fig. 2.3 Giant Túró Rudi

References

Britz TJ, Robinson RK (2008) Advanced dairy science and technology. Blackwell, Oxford
Dairy Handbook. Alfa-Laval Food Engineering, Sweden
Fenyvessy J, Csanádi J (2007) Dairy technology (Tejipari technológia), Universitas Szeged
Szakály S (Ed) (2001) Dairy technology and economy (Tejgazdaságtan). Dinasztia, Budapest
    ISBN 963 657 333 6
Walstra P, Wouters JTM, Geurts TJ (2006) Dairy science and technology. Taylor & Francis, Boca
    Raton

Chapter 3
Quark: A Traditional German Fresh Cheese

S. Drusch and U. Einhorn-Stoll

Contents

3.1 Introduction

Quark is a traditional German unripened cheese. Similar products are known in other countries, e.g., cream cheese, Topfen, cottage cheese, tvorog, and fromage frais. Unripened fresh cheese nowadays is still very popular in European and Northern American countries; the production of fresh cheese and quark in Germany in 2013 alone amounted to 842.000 t (Statista 2014).

According to the Codex Alimentarius General Standard for Cheese, cheese is generally defined as a ripened or unripened soft, semihard, hard, or extra-hard product, which may be coated, and in which the whey protein/casein ratio does not exceed that of milk.

S. Drusch (✉) • U. Einhorn-Stoll
Institute of Food Technology and Food Chemistry, Technische Universität Berlin,
Berlin, Germany
e-mail: stephan.drusch@tu-berlin.de

© Springer Science+Business Media New York 2016   21
A. McElhatton, M.M. El Idrissi (eds.), *Modernization of Traditional Food*
Processes and Products, Integrating Food Science and Engineering Knowledge
Into the Food Chain 11, DOI 10.1007/978-1-4899-7671-0_3

The product is obtained by:

(a) coagulating wholly or partly the protein of milk, skimmed milk, partly skimmed milk, cream, whey cream or buttermilk, or any combination of these materials, through the action of rennet or other suitable coagulating agents, and by partially draining the whey resulting from the coagulation, while respecting the principle that cheese-making results in a concentration of milk protein (in particular, the casein portion), and that consequently, the protein content of the cheese will be distinctly higher than the protein level of the blend of the above milk materials from which the cheese was made and/or

(b) processing techniques involving coagulation of the protein of milk and/or products obtained from milk, which give an end product with similar physical, chemical, and organoleptic characteristics as the product defined under (a).

Unripened cheese including fresh cheese is cheese which is ready for consumption shortly after manufacture (Codex Alimentarius Commission 2008a, b). The composition of unripened cheese is defined in a dedicated standard, the Codex Group Standard for unripened cheese including fresh cheese (Codex Alimentarius Commission 2008b). Based on this standard only the following ingredients are allowed in the production of fresh cheese:

– Starter cultures of harmless lactic acid and/or flavor producing bacteria and cultures of other harmless microorganisms
– Rennet or other safe and suitable coagulating enzymes
– Sodium chloride
– Potable water
– Gelatin and starches: used in the same function as stabilizers
– Vinegar
– Rice, corn and potato flours and starches as anticaking agents for treatment of the surface of cut, sliced, and shredded products

Specific German criteria for the ingredients of quark and its composition are laid down in a German Regulation on Cheese (Käseverordnung 1985). According to this regulation, milk, cream, skim milk, or whey from the process of quark production shall be used as ingredients. Table 3.1 summarizes specific criteria for different types of quark. The whey protein content shall not exceed 18.5 % of the total protein content.

The typical chemical composition of quark is summarized in Table 3.2.

Concerning the sensory properties, quark should have a color from milky white to creamy yellow (depending on the fat content), a uniform smooth texture and a light, acidic taste and smell without any noticeable off-flavor.

Table 3.1 Requirements for different types of cheese including quark (Käseverordnung 1985)

Class	Fat content [% dry matter]	Minimum content of dry matter [%]	Minimum protein content [%]
Magerstufe (low fat)	<10	18	12.0
Viertelfettstufe (quarter fat)	>10	19	11.3
Halbfettstufe (half fat)	>20	20	10.5
Dreiviertelfettstufe (three-quarter fat)	>30	22	9.7
Fettstufe (fat)	>40	24	8.7
Vollfettstufe (full fat)	>45	25	8.2
Rahmstufe (cream)	>50	27	8.0
Doppelrahmstufe (double cream)	60–85	30	6.8

Table 3.2 Chemical composition of different types of quark (Bundesministerium für Landwirtschaft, Ernährung und Verbraucherschutz 2010)

	Low fat	Half fat	Fat
Energy (kcal/100 g)	75.3	109.5	158.7
Protein (g/100 g)	13.5	12.5	11.1
Fat (g/100 g)	0.2	5.1	11.4
Carbohydrate (g/100 g)	4.0	2.7	2.6
Water (g/100 g)	80.7	78.1	73.4
Calcium (mg/100 g)	114	85	95

3.2   Basics of Quark Processing

Quark traditionally consists of acidic precipitated caseins, the major milk protein fraction (80 %). The precipitation takes place at the isoelectric point of the caseins at pH about 4.6; whey proteins (20 % of the milk proteins) do not precipitate. The acid for precipitation is produced by lactic acid bacteria, naturally present or added to the product. They metabolize the lactose to lactic acid. Nowadays, lactic acid bacteria are chosen as starter cultures according to their ability to produce lactic acid and special aroma compounds and regarding their phage resistance. Upon acidification, calcium, that stabilizes the casein micelle, is solubilized from calcium phosphate bridges and the micelles themselves disintegrate. The dissolved calcium is removed in the process later on; therefore, the calcium content of quark is rather low compared to rennet-coagulated cheese. Furthermore, the surface charge of the casein micelles decreases and the caseins aggregate due to hydrophobic interactions. The supernatant after precipitation, the whey, containing whey proteins, calcium and water, can be removed from the precipitate in different ways.

Apart from the pure acid coagulation ("Sauermilchquark") also a combined acid/rennet coagulation process ("Labquark") can be performed. Rennet is a mixture of proteolytic enzymes, which originally was extracted from the stomach mucosa of young sucking calves. The most important enzyme is the chymosin. As a result of the increased consumption, the demand for rennet increased considerably and classical calf rennet had to be replaced. Extracts from special plants ("Labkraut"; *Galium verum*) gave no sufficient results (Teichert 1931) and also enzymes from adult cattle or pigs such as pepsin had side effects like bitter taste. Therefore, nowadays rennet is produced with biotechnological methods in high amounts in a constant quality. The primary cleavage site for chymosin is a specific bond between two amino acids (Phe_{105}-Met_{106}) of κ-casein, an amphiphilic molecule that is stabilizing the casein micelles in the aqueous environment. However, also proteolytic activity against the other caseins, e.g., β-casein at Leu_{192}-Tyr_{193}, has been reported (Fox et al. 2000). As reviewed by Schulz-Collins and Senge (2004), combined aggregation by renneting and acidification leads to a synergistic effect with acidification potentiating the aggregating tendency of the rennet-treated casein particles.

3.3   Traditional Quark Processing

The production of quark developed from the naturally occurring acidification during storage of milk and subsequent coagulation of the milk proteins. The first quark was made more or less "by accident": In order to increase the yield of cream, milk was stored for a prolonged period, during which natural acidification occurred. The resulting product was found to be useful as a food. In the nineteenth century, quark was almost exclusively produced for home requirements. The main focus on the farm and in small dairy manufacturers was, for economic reasons, on the production of butter (Kalka 2004). A milk product booklet from 1911 does not even mention quark as an industrial product (Reitz 1911). Nevertheless, use of homemade quark is described in traditional cooking and household books of that time.

At the end of the nineteenth century, industrial milk processing developed and later on quark production from skim milk as a by-product of cream separation on a commercial scale started. Knoch (1930) described quark as a fresh by-product, which is made in the case of excess milk supply in a dairy company. Sometimes, quark was produced also from slightly acidified raw milk that could not be processed to fresh drinking milk any more (Teichert 1931). The main problem of quark as a commercial food product at that time was the stability during transport and storage without permanent cooling. Therefore, in an overview of commercial food products ("Warenkunde") from 1926 Quark is not mentioned as a regular food product but only as an intermediate product for some cheese types (Kahrs 1926).

A prolongation of shelf-life through implementation of a more hygienic manufacturing practice and refrigerators in households was the driving force behind an industrial quark production development. Henkel (1933) described that quark for direct consumption ("Speisequark") was produced by addition of lactic acid bacteria

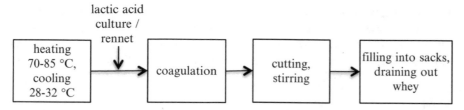

Fig. 3.1 Traditional production of quark around 1930

Fig. 3.2 Traditional quark
processing equipment.
(a) Pressing machine
("Quarkpresse") for quark
sacks; (b) Quarkfertiger—
vessel with perforated lid;
(c) "Passiermaschine";
(d) Filling machine
(reproduced with
permission from Niemeyer
1951)

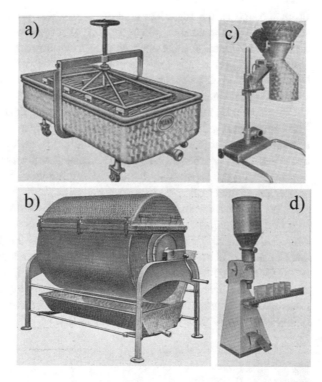

to the milk and according to Teichert (1931) also the combined acid/rennet coagulation process was applied in an industrial scale as shown in Fig. 3.1. Quark is mentioned also in a new list of commercial dairy products (Schischke et al. 1934).

The coagulation took place in big vessels ("Quarkwanne"). The traditional method of draining out the whey was to fill quark into linen or cotton sacks and to form piles of these sacks, causing pressure on the lower layers. After 7–9 h and several re-packings of the sacks, they were emptied and cleaned. For a more defined pressure, the sacks were also treated in a pressing machine (Fig. 3.2a). Later on, coagulation was performed in a special vessel ("Quarkfertiger") with a perforated lid. After fermentation, the coagulum was moderately cut and covered with a cloth, the vessel was turned for 180° and the whey could drain out through the lid (Fig. 3.2b). The final water content should always be below 75 % (Henkel 1933);

therefore, the resulting quark was rather dry and crumbly. For a better consistency, the quark could be treated in a special apparatus ("Passiermaschine", Fig. 3.2c), where also the addition of cream for special purposes was possible. Finally, the quark was filled into barrels or, for more convenience, into small packages (Fig. 3.2d).

3.4   Classical Industrial Quark Processing

From the mid of the twentieth century, quark was produced in a large industrial scale. This was forced by a new technology for removing the whey. Quark separation by centrifugation developed as it is still performed nowadays. This process considerably reduced the time for draining out the whey and was more hygienic because no clothes or sacks were necessary any more. Moreover, a constant but variable water content and quality could be achieved.

The separation step is performed in a specifically designed separator (Fig. 3.3). The coagulum is axially fed into the separator and distributed into the disk stack in the separating chamber via the rising channels. The disk stack facilitates separation and mass flow within the chamber. The coagulated protein, the quark, is distally removed from the separating chamber, while the whey is removed centrally in the upper part of the bowl.

A typical classical industrial process for quark production is shown in Fig. 3.4. In a first step the milk is pasteurized in order to inactivate pathogenic microorganisms. This is usually performed at 72–75 °C for 15–30 s. The milk is cooled down and the lactic acid bacteria are added to the coagulation tank. After stirring, the milk is left for approximately 90 min until the pH reaches approximately 6.3. Rennet is added and acidification continues until a pH of 4.5 is reached and a coagulum has developed. The coagulum is thoroughly stirred to create a homogenous structure (GEA Westfalia Separator Group GmbH 2010).

Fig. 3.3 Schematic drawing of a QUARK separator (Courtesy of GEA Westfalia Separator Group GmbH, with permission)

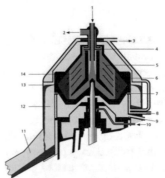

1 Feed
2 Discharge, whey
3 Discharge / cover cooling
4 Centripetal pump, whey
5 Disc stack
6 Segmental insert
7 Brake ring, cooled
8 Feed, concentrate collector and brake ring
9 Discharge, frame
10 Sterile air / CIP connection
11 Discharge to quark hopper
12 Concentrate collector
13 Nozzles
14 Rising channels

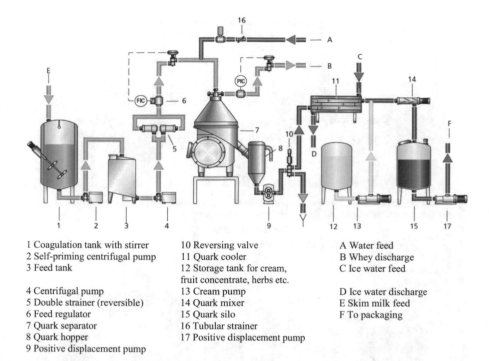

1 Coagulation tank with stirrer
2 Self-priming centrifugal pump
3 Feed tank

4 Centrifugal pump
5 Double strainer (reversible)
6 Feed regulator
7 Quark separator
8 Quark hopper
9 Positive displacement pump

10 Reversing valve
11 Quark cooler
12 Storage tank for cream, fruit concentrate, herbs etc.
13 Cream pump
14 Quark mixer
15 Quark silo
16 Tubular strainer
17 Positive displacement pump

A Water feed
B Whey discharge
C Ice water feed

D Ice water discharge
E Skim milk feed
F To packaging

Fig. 3.4 Processing line for quark using a standard process (Courtesy of GEA Westfalia Separator Group GmbH, with permission)

3.5   Modern Quark Processing

Though the classical process is still applied in smaller and more traditional dairy factories and sometimes also for special products, an alternative process with the major advantage of an increase in quark yield has been developed and is nowadays widely used in big companies of the dairy industry. The process is called "Thermo-Quark" process and includes an intense heating step at the beginning (Fig. 3.5). After cream separation, skim milk is heated at 82–88 °C for 5–6 min. The temperature is well above the temperature required for the denaturation of the whey proteins in milk. During denaturation free thiol groups become available and undergo a covalent binding via disulfide bridges with the caseins. As a consequence, the whey proteins are bound and remain in the quark. After thermal treatment, the milk is cooled down and acidification and renneting is performed in a similar manner to the conventional process. Since covalent binding of the whey proteins to casein sterically hinders the enzymatic cleavage, the amount of rennet is higher compared to the conventional process. Acidification to the final pH of 4.5 needs approximately 16 h (GEA Westfalia Separator Group GmbH 2010). In order to facilitate separation in the quark separator, the coagulum is pre-heated to 40–45 °C by blending thermally treated coagulum (60–65 °C, 2–4 min) with non-heated coagulum.

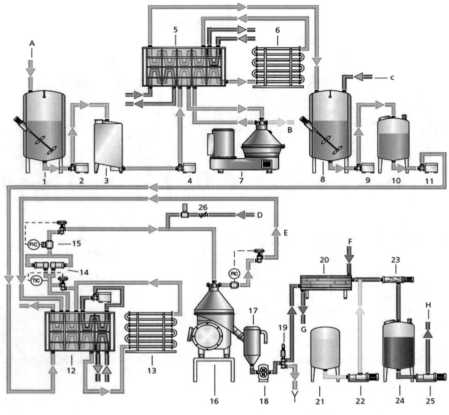

1 Raw milk silo with stirrer	12 Plate heat exchanger	24 Quark silo
2 Centrifugal pump	13 Tubular regenerator	25 Positive displacement pump
3 Balance tank	14 Double strainer (reversible)	26 Tubular strainer
4 Centrifugal pump	15 Feed regulator	A Raw milk feed
5 Plate heat exchanger	16 Quark separator	B Cream discharge
6 Tubular regenerator	17 Quark hopper	C Culture and rennet dosing
7 Skimming separator	18 Positive displacement pump	D Water feed
8 Coagulation tank for vat milk with stirrer	19 Reversing valve	E Whey discharge
9 Centrifugal pump	20 Quark cooler	
10 Balance tank with stirrer	21 Cream tank	F Ice water feed
11 Centrifugal pump	22 Positive displacement pump	G Ice water discharge
	23 Quark mixer	H To packaging

Fig. 3.5 Processing line for quark using the thermo-quark process (Courtesy of GEA Westfalia Separator Group GmbH, with permission)

A new technological trend for some non-ripening cheese products since the 1980s is the application of membrane technology (GmbH 2014). There are several possibilities for the implementation of membrane processing steps (Fig. 3.6).

1. A part of the milk can be treated by ultrafiltration or microfiltration before addition of bacteria and rennet in order to achieve a certain increase of dry matter (pre-concentration) up to 4 %. After fermentation, the whey is drained off by separation as described above.

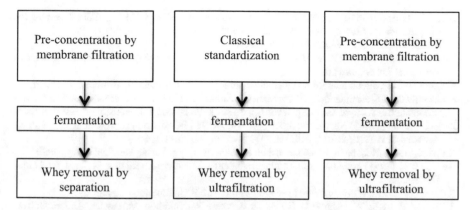

Fig. 3.6 Possible implementation of membrane filtration steps into quark processing

2. The process can be similar to the thermo-quark process as described above, only the whey is removed using ultrafiltration.
3. In a combined process pre-concentration as well as whey separation is performed by ultrafiltration.

Nowadays, commercial quark is a protein-rich smooth product with a creamy texture, nearly independent on the fat content. Quark is a common ingredient in dairy-based and bakery products but is consumed also in a more or less pure form. A broad variety of commercial sweet quark desserts or flavored spreads containing herbs and other spices is available and even ice-cream has been produced from quark. Low-fat quark is helpful in diets but also for cooking as a thickener in sauces and dressings. Quark can be consumed as a plain product on bread or as a dip but is an ingredient in different sweet and savoury dishes. However, it is important to know that the application of traditional recipes for homemade food from old cook books, such as quark cake or soufflé, can be difficult because of the completely different texture properties of the traditional quark in comparison to the modern industrial product.

References

Codex Alimentarius Commission (2008a) Codex Group Standard for Unripened Cheese including Fresh Cheese. Codex Standard 221-2001. http://www.codexalimentarius.net/download/standards/363/CXS_221e.pdf. Visited 20 May 2011

Codex Alimentarius Commission (2008b) Codex General Standard for Cheese. Codex Standard 283-1978. http://www.codexalimentarius.net/download/standards/175/CXS_283e.pdf. Visited 20 May 2011

Bundesministerium für Ernährung, Landwirtschaft und Verbraucherschutz (2010) Bundeslebens mittelschlüssel. Version 3.01

Fox PF, Guineem TP, Cogan TM, McSweeney PLH (2000) Fundamentals of cheese science. Aspen, Gaithersburg

GEA Westfalia Separator Group GmbH (2010) Process lines from GEA Westfalia Separator for the production of soft cheese. 9997-8254-030/0810 EN

GEA TDS GmbH (2014) Fermented fresh cheese

Henkel E (1933) Die Milch. In: Dammer O (ed) Chemische Technologie der Neuzeit. Verlag von Ferdinand Enke, Stuttgart

Kahrs F (1926) Warenkunde für den Kolonialwaren- und Feinkosthandel. Weltbund, Hamburg

Kalka E (2004) Lebensmittelqualität zwischen Geschmack und Zeichen—Eine natur- und kulturwissenschaftliche Analyse am Beispiel von Speisequark. Ph.D. thesis. University of Kassel, Kassel

Käseverordnung (1985) Neugefasst durch B. V. 14.04.1986 BGBl. I S. 412; zuletzt geändert durch Artikel 4 V. v. 17.12.2010 BGBl. I S. 2132; Geltung ab 23.11.1985

Knoch C (1930) Handbuch der neuzeitlichen Milchverwertung. Verlag von Paul Parey, Berlin

Niemeyer H (1951) Handbuch für Molkereifachleute. Milchwirtschaftlicher Verlag Karl Mann, Hildesheim

Reitz A (1911) Die Milch und ihre Produkte. Verlag von B.G. Teubner, Leipzig

Schischke F, Mohr W, Fehlow G (1934) Dr. Oetkers Warenkunde, Verlag Dr. August Oetker, Nährmittelfabrik Bielefeld

Schulz-Collins, D, Senge, B (2004) Acid- and acid/rennet-curd cheeses part A: quark, cream cheese and related varieties. In Fox PF, Mc Sweeney PLH, Cogan, TM, Guinee, TP (eds) Cheese Chemistry, physics and microbiology, 3rd ed, vol.2 Major chees groups p. 301-328. Elsevier Academic Press, San Diego, CA

Statista (2014) Produktion von Frischkäse in Deutschland in den Jahren 1999 bis 2013. http://de.statista.com/statistik/daten/studie/5579/umfrage/produktion-von-frischkaese-in-deutschland-seit-1999/. Visited 23 Sept 2014

Teichert K (1931) Deutsches Käsereibuch. Verlagsbuchhandlung von Eugen Ulmer, Stuttgart

Chapter 4
Modernization of the Traditional Irish Cream Liqueur Production Process

Peter C. Mitchell

Contents

4.1 Introduction

The spirit drinks sector in Ireland has a long tradition and strong links to the agricultural sector. Whilst whiskey is the spirit drinks category with the greatest heritage in Ireland, it is the cream-based liqueur category which has grown to prominence in Irish food history. Today, Baileys as the number one selling liqueur brand in the world leads the 126 million litre global cream-based liqueur category, with a 46 %

P.C. Mitchell (✉)
Northern Ireland Centre for Food and Health, University of Ulster,
Coleraine, Northern Ireland, UK
e-mail: pc.mitchell@ulster.ac.uk

© Springer Science+Business Media New York 2016
A. McElhatton, M.M. El Idrissi (eds.), *Modernization of Traditional Food Processes and Products*, Integrating Food Science and Engineering Knowledge Into the Food Chain 11, DOI 10.1007/978-1-4899-7671-0_4

volume share (Cunnington 2010). The rest of the category is made up of a number of small global brands and small regional/local producers (Cunnington 2010).

Varnam and Sutherland (1994a) state that liqueurs may be defined as distilled spirits, which have been sweetened and flavoured with substances of compatible taste. Traditional methods and practices prevailed in the manufacture of many liqueurs, until their classification became better defined, particularly in Europe and the USA. As a sub-classification of spirit drinks, a liqueur must contain at least 15 % alcohol by volume (Clutton 2003). However, there are significant differences in the liqueur definitions within Regulation (EC) No 110/2008 of the European Parliament (European Commission 2008) and the Federal Alcohol Administration Act (Alcohol and Tobacco Tax and Trade Bureau 2007), particularly in terms of minimum sugar content. However, the definitions still allow for a wide variety of liqueurs to be produced.

Commercial cream liqueurs are added value, long-life, oil-in-water emulsions, combining the flavour of an alcoholic drink with the texture of thickened cream. To control the kinetics of the processes that lead to the breakdown of cream liqueurs requires a device to break up the milk fat globules and disperse them within the emulsion system, and the addition of stabilizing chemical additives (Bergenstahl 1995) to reduce the interfacial tension and prevent flocculation and coalescence of the dispersed phase (Kinsella and Whitehead 1989). The dispersed phase is more likely than not to be partially solidified, and to an extent which is quite strongly temperature dependent, and thus reference is made interchangeably within this review to dispersed phase droplets and particles. The high-pressure homogenizer and caseinates are the respective widely used device and stabilizer in the production of traditional Irish Cream liqueurs. The inclusion of caseinate means that the product cannot be consumed with a mixer drink, as the resulting pH drop would cause protein precipitation. Traditional Irish Cream liqueurs are some of the finest food emulsions (Dickinson and Stainsby 1988) where an exemplar manufacturing specification is 90 % of particles less than 0.5 μm. The particle size of the dispersed phase droplets influences not only storage stability but also the palatability, mouthfeel, texture and general appearance of traditional Irish Cream liqueurs.

Whilst the technology platform for the manufacture of stable cream liqueurs has been in place since 1980s (Banks and Muir 1988), the commercial importance of these products continued to drive scientific and technical investigations aimed at increasing volume of production and range of products, and enhancing product quality in a more demanding global supply chain. Today the use of alternative sources (Medina Torres, Calderas, Gallegos-Infante, González-Laredo, and Rocha-Guzmán 2009) and new types of ingredients, and enhanced process technologies (Heffernan, Kelly, and Mulvihill 2009) are under investigation in order to lower the cost of manufacture and formulate novel products. So in just over 35 years, traditional Irish Cream liqueur products and production processes have evolved and modernized. This literature review will focus on the problems encountered and solved, whilst emphasizing the control of homogenization parameters, in the production of stable, high-quality traditional cream-based liqueurs.

Table 4.1 Composition of a traditional cream liqueur

Component	Traditional product example
Milk fat	16 % w/w
Added caseinate	3.3 % w/w
Added sugar	19 % w/w
Ethanol	14 % w/w[a]
Total solids	40 % w/w
Fat to protein	4.2

[a]Corresponds to 17 % v/v

4.2 Irish Cream Liqueurs

The composition of cream liqueurs can vary widely with milk fat, added caseinate, added sugar and ethanol content ranging from 4.5 to 16.5 % w/w, 1.8 to 3.8 % w/w, 17.5 to 22 % w/w and 12.7 to 15.4 % w/w, respectively (Banks et al. 1981a). The traditional Irish Cream liqueur is a premium product, containing whiskey as part of the spirit content and a composition which is at high end of the milk fat range. The composition of one commercial traditional Irish Cream liqueur is shown in Table 4.1.

The manufacture of cream liqueurs is detailed by Banks et al. (1981a), and Banks and Muir (1985, 1988). The two methods of manufacture discussed are the single-stage and two-stage methods. In the former, caseinate and sugar are dissolved in the water at 85 °C by vigorous stirring, and the air is dispersed from the mixture. Next double cream is blended into the mixture, and alcohol is added. This pre-emulsion is then heated to 55 °C and homogenized at 27.6 MPa (4000 psi). The two-stage method differs in that alcohol is added after homogenization. An example of a commercial single-stage process flow diagram for the production of traditional Irish Cream liqueurs is shown in Fig. 4.1. This highlights the importance of homogenization parameters, including the use of the two-stage valve and multiple passes, which will be covered in Sect. 4.3.

The formulation, testing and stability of 16 % fat by weight cream liqueurs have been researched (Power 1996). Cream, ethanol stability, the defects associated with cream liqueurs and the key controls to prevent and minimize destabilizing mechanisms merit particular attention.

4.2.1 Cream

Cream is produced from unhomogenized milk by mechanical separation of the fat globules. Cream is an oil-in-water emulsion where the globules are dispersed in an aqueous (serum) phase consisting of protein, lactose and salts. Cream is defined in the UK as "that part of milk rich in fat that has been separated by skimming or otherwise", and legally contains not less than 12 % fat by weight (Early 1998).

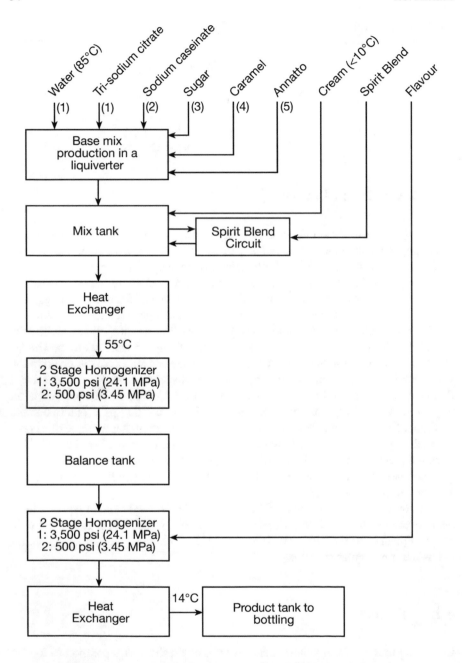

(1), (2), (3), (4) and (5) denote the sequence of addition of ingredients
in base mix production, where the mix rests for 15 minutes between
(2) and (3)

Fig. 4.1 Example of a commercial single-stage process flow diagram for the production of traditional Irish Cream liqueurs

The World Health Organization standards for the minimum fat content by weight of creams are: half, 10–18 %; single, 18 %; whipping, 28 %; heavy whipping, 35 % and double, 45 % (Varnam and Sutherland 1994b).

Cream suitability for processing relates to fat content, acidity, seasonality, storage, handling and heat treatment (Rothwell 1989; Towler 1994). Cream homogenization and increased storage time increase cream viscosity. Winter cream has poorer ethanol stability in comparison to cream separated from mid-lactation milk (Rothwell 1989).

Double cream is the most usual source of milk fat in traditional Irish Cream liqueurs. Double cream contains surface active materials at *circa* 0.55 % by weight, which includes protein and the naturally occurring milk fat globule membrane (Anderson 1991). The protein to fat ratio in double cream is *circa* 0.04 (Muir and Banks 1986). When double cream undergoes slight homogenization, fat globules clump together and cluster, and there is a marked thickening of the cream (Mulder and Walstra 1974). Cream liqueurs employ a much higher protein to fat ratio, and also use more severe homogenization pressures (Muir and Banks 1986). It is the use of double cream at *circa* 33 % of the formulation that gives the traditional Irish Cream liqueur its richness and body, and masks the harshness of the alcohol (Muir and Banks 1984).

4.2.2   Ethanol Stability

The addition of ethanol is used as an indicator of milk stability. The higher the calcium concentration, the less stable the caseinate complex to ethanol (Davies and White 1958). Horne and Parker (1980) increased the ethanol stability of milk by increasing the pH above 7. Horne and Muir (1990) demonstrated that the level of free calcium controls the ethanol stability of the milk system. This finding has proved extremely useful in the production of cream liqueurs. The inclusion of citrate to sequester calcium prevents calcium-associated instability eliminating one of the major defects of cream liqueurs (Banks et al. 1981a).

4.2.3   Defects Associated with Cream Liqueurs

The main defects which occur in cream liqueurs are: (1) creaming, where the fat forms a layer at the surface of the liqueur; (2) formation of a plug of cream or fat in the neck of the bottle; (3) gelation of the bulk of the product with some accompanying syneresis and (4) appearance of a slight granular precipitate at the bottom of the bottle (Banks and Muir 1985; Banks and Muir 1988; and Dickinson, Narhan, and Stainsby 1989).

Creaming is the gravitational separation of emulsified droplets. This results in the formation of layers, with oil volume fractions higher than that of the original emulsion. The creamed layer is usually visible to the naked eye.

Creaming in cream liqueurs is usually reversible as gentle shaking will redisperse the fat layer. The formation of a cream plug or "cohesive cream" in cream liqueur is more likely in cream liqueur formulations at low pH, high calcium content or low emulsifier content. The effect of these factors will be increased substantially by temperature fluctuations during storage (Banks et al. 1981a; Dickinson et al. 1989).

The type of gel which may form due to instability in cream liqueurs is classified as a particle gel (Clark 1992; Power 1996). This gel undergoes syneresis with the separation of a clear liquid at the bottom of the bottle.

The sodium or potassium salts of citric acid may be used to sequester calcium in emulsions. Ionic calcium has a major effect on the stability of dairy products including cream liqueurs (Banks et al. 1981a; Davies and White 1958; Horne and Muir 1990). Citrate binds with the calcium, preventing instability. The appearance of a slightly granular precipitate has occasionally been observed at the bottom of the bottle. This is believed to be excess citrate. Citrate complexes the calcium in the liqueur and reduces the possibility of gelation and thus increases the shelf life at ambient temperature from months to years (Banks and Muir 1988). In the traditional product example shown in Table 4.1, tri-sodium citrate di-hydrate is used at a level of 0.19 % by weight of the total formulation.

4.2.4   Controls to Prevent and Minimize Destabilization of Cream Liqueurs

Steric (polymeric) repulsion, electrical double layers (electrostatic repulsion) and increased continuous phase viscosity are the main mechanisms for the stabilization of food emulsions (Schubert and Armbruster 1989). Steric repulsion, attributed to the protruding part of the casein molecule, and the repulsion by charged protrusions of the particles with similar charges are very important in the formation of stable cream liqueurs (Narhan 1987). As the alcohol level increases in a cream liqueur, there is a reduction in steric repulsion and an increase in the viscosity of the continuous phase, with a resultant increase in particle aggregation and coalescence (Banks and Muir 1985). However, alcohol-induced aggregation will occur above 17.5 % w/w alcohol strength, which must be avoided (Banks and Muir 1988). Excess heat treatment of cream must also be avoided in the production of traditional Irish Cream liqueurs so as to prevent aggregation by cross-linking (Power 1996). The addition of tri-sodium citrate to bind ionic calcium, which would otherwise aggregate casein micelles, and ensuring the pH is above 6.8 are also important controls in the production of traditional Irish Cream liqueurs (Muir 1989). Temperature controls are also important so as to minimize fat crystallization, which could result in bridging between particles, and to avoid temperature cycling with agitation, which could result in the desorption of the protective protein layer (Narhan 1987).

4.3   Homogenization Parameters

In cream liqueur production, the term "homogenization" and "homogenizer" refer to the process and equipment related to the classically recognized homogenizer, which was first developed by Auguste Gaulin for milk treatment (Tunick 2009). In such a high-pressure homogenizer, a positive displacement pump, usually of plunger or piston type, is used to force the liquid into the homogenizing valve where mechanical energy is used to break drops into smaller droplets (Wilbey 1992).

The premix enters the valve seat at a relatively constant rate of flow, relatively low velocity and high pressure (10–40 MPa or 1450–5801 psi) and, as it begins to move into the narrow adjustable gap between the valve and the seat (15–300 μm), it undergoes a very rapid rise in velocity and decrease in pressure causing turbulence, cavitation and intense mixing. For example, a pressure drop of 13.8 MPa or 2000 psi causes a velocity in excess of 160 m/s and the whole process of homogenization is complete in less than 50 μs (Pandolfe and Kinney 1983). Turbulence on the low-pressure side of the valve is probably the most important factor leading to the formation of fine droplets (Dickinson 1992). Whilst the correlations between characteristics of turbulence and cavitation, parameters of homogenizing valves and homogenization efficiency are insufficiently known (Rovinsky 1994), much is now understood about the efficient operation of high-pressure homogenization in Irish Cream liqueur production. Current understanding of homogenization parameters is now briefly discussed under the following headings: formulation; premix quality; valve technology; homogenization process; and troubleshooting protocol.

4.3.1   Formulation

Important characteristics of formulation which influence homogenization efficiency are the amount and type of emulsifier, the dispersed phase concentration and the viscosity of the dispersed and continuous phases.

The emulsifier lowers the interfacial tension between the two phases being homogenized, and therefore makes more efficient use of the available energy. Less homogenizing energy is required to overcome interfacial tension. Also, the emulsifier stabilizes the new interfacial area formed during homogenization and later prevents coalescence and agglomeration of the droplets. The cost of emulsifier is many times more than the cost of the mechanical energy thus the high-energy homogenizer is more economical than the low-energy mixer for emulsification (APV n.d.c). If the concentration of the emulsifier is insufficient or the wrong type of emulsifier is used, overworking occurs (Pandolfe 1995). This can be recognized if an increased average droplet size or even the formation of two separate phases of the emulsion is obtained on increasing the homogenization pressure. This is because the emulsifier cannot cover the expanded surface of the smaller droplets leading to a rate of coalescence which exceeds the rate of disruption. At a low protein to fat ratio of 0.10,

Muir and Banks (1986) found that the long-term stability of a traditional cream liqueur, which used sodium caseinate as the stabilizer, was poor, and when such a premix was subject to severe homogenization, it could lead to the formation of large particles with subsequent creaming and fat plug formation. Whilst the protein to fat ratio is 0.21 in the traditional Irish Cream liqueur example (Table 4.1), there are other stable 16 % w/w fat and 17 %v/v traditional Irish Cream liqueurs which use a lower protein to fat ratio by combining a low molecular weight emulsifier, such as the nonionic glycerol monostearate (Euston 2008), with a reduced level of sodium caseinate in the premix. In the production of cream liqueurs there are advantages to be gained from using certain types of caseinate (Muir and Dalgleish 1987), and thus manufacturers have a preferred caseinate supplier. Lynch and Mulvihill (1997) concluded that changes in the apparent viscosity of cream liqueurs on storage at 45 °C are caseinate dependent, and suggested that electrostatic and sulphydryl interactions may be involved in these changes. Medina Torres et al. (2009) also observed significant changes, with storage time at 40 °C, in apparent viscosity and particle size distribution between caseinate batches which differed in metallic ion content. The best caseinate batch for longer term cream liqueur stability and reduced age thickening had the lowest total metallic (Ca^{++} and Na^+) ion content.

As the concentration of the dispersed phase is increased, the probability of the droplets not being reduced in size is increased (Pandolfe 1995). At a constant fat to emulsifier ratio and constant homogenization pressure, the increase in average particle size with increased dispersed phase concentration from 4.5 to 16.5 % w/w milk fat would be very small. However, once above a minimum milk fat volume fraction which is homogenization pressure dependent, the average particle size will increase with increasing dispersed phase concentration (Phipps 1983). When the emulsifier is required to handle an increasing surface area, it may not be able to stabilize the interface before droplets collide and coalescence occurs.

As the dispersed phase viscosity is increased, the average particle size of the homogenized product decreases. As the continuous phase viscosity is increased, there is a slight increase in the average particle size of the homogenized product, but above a viscosity of 50 mPa.s (cP), the average particle size remains constant (Pandolfe and Kinney 1983). The specification for apparent viscosity at 25 °C and 39.6 s^{-1} shear rate for the traditional Irish Cream liqueur example given in Table 4.1 is 28–38 mPa.s (cP).

4.3.2   Premix Quality

At any given homogenizing pressure, a fixed amount of energy is available for particle size reduction. If a significant portion of this energy is needed to reduce very large particles, then there will not be enough energy left over to work on the smaller particles and reduce them even further (Masucci 1989a). Thus to maximize homogenization efficiency, it is important to improve premix quality, as measured by uniformity and smaller particle size, whether by altering raw material quality, the formulation, the batch preparation method or the mechanical premix device employed (Masucci 1989a; Pandolfe and Kinney 1983). In cream liqueur production, the premixing of phases to form a pre-emulsion must be preceded by the dispersion of the powdered ingredients. Caseinates are very cohesive and have a tendency to form agglomerates. Caseinates are difficult to dissolve and will rapidly increase in apparent viscosity, especially if added directly to cream (Silverson n.d.). In cream liqueur production, long mixing times, the resting of caseinate after dissolution and the testing of the pre-emulsion prior to homogenization for % of particles less than 2 µm and apparent viscosity (39.6 s^{-1} shear rate) at the temperature at which homogenization occurs (55 °C), have traditionally been used to ensure an acceptable premix quality. In recent times, the addition of a high-shear in-line mixer to the existing cream liqueur production process has been adopted by one manufacturer so as to give agglomeration-free dispersion of caseinate, rapid mixing times, a more uniform, stable pre-emulsion of low particle size and faster processing through the high-pressure homogenizer (Silverson n.d.). Short- and long-term stability benefits have been realized in cream liqueur production practice, although use of the in-line mixer can lead to greater age thickening of the cream liqueur.

Inclusion of air in the raw product, during premixing but more frequently by leakage through the seals and/or gaskets of the homogenizer system, reduces homogenization efficiency and results in the formation of a cream line shortly after homogenization as very small fat particles which would not have separated otherwise adhere to the surfaces of the air bubbles which escape to the surface of the product. Increasing the homogenizing pressure will result in even smaller bubbles of air; therefore, an even more severe separation of the fat occurs as these air bubbles escape to the surface (APV, n.d.a).

4.3.3 Valve Technology

The key to any homogenizer is valve technology. Whilst there a wide variety of high-technology valves for different applications (Pandolfe 1999), a standard "plug-flow" valve geometry configured as a two-stage valve assembly is used for most emulsions, including cream liqueurs. An exception in dairy emulsions is the use of a Gaulin Micro-Gap homogenizing valve assembly in high volume liquid milk processing to produce the same or better product quality at up to 40 % lower pressure, and thus reduced energy cost, and with increased valve life (Pandolfe 1999).

In a two-stage design, the second stage establishes controlled backpressure, concentrating the energy in the first homogenizing zone and also minimizing the possibility of clumping/coalescence. In evaluating valve combinations, it has been found that with fluid milk at any given homogenizing pressure, efficiency is increased by the use of a second-stage valve, whereby 10 % to a maximum of 20 % of the total pressure is applied by the second-stage valve (APV, n.d.b). To give the desired total pressure, which can be 27.6 MPa (4000 psi) in cream liqueur production, the second-stage valve is set first, at for example 12.5 % of the total pressure (3.45 MPa; 500 psi), followed by the first-stage valve, to give an actual pressure drop of 24.1 MPa (3500 psi) through the first-stage valve.

4.3.4   Homogenization Process

In the homogenizing valve, according to turbulence theory it has been shown that the homogenizing pressure (P) is related to the average volume/surface diameter (d_{vs}) by Eq. (4.1)

$$d_{vs} \text{ is proportional to } P^{-0.6} \tag{4.1}$$

So increasing the homogenizing pressure will reduce the average particle size and increase the energy input costs (Pandolfe and Kinney 1983). Banks et al. (1981b) reported the typical total pressure for homogenization in pilot plant cream liqueur production as 27.6 MPa (4000 psi), but with a requirement for a second pass through the homogenizer at the same total pressure. Muir and Banks (1986) conducted their pilot scale work on multiple homogenizations of cream liqueurs at 31.0 MPa (4500 psi). Widmar and Tripp (1990) presented their single pass process, using a total pressure for homogenization of 34.5 MPa (5000 psi), for the preparation of cream-based liqueurs.

The temperature of homogenization is important, particularly as it relates to the viscosity of the dispersed phase. Banks et al. (1981b) used a homogenization temperature of 55 °C in the two pass pilot scale production of cream liqueurs at a total homogenization pressure of 27.6 MPa (4000 psi), and subsequently Muir and Banks (1986) used 50 °C in a multiple homogenization of cream liqueurs investigation.

Many applications require a very uniform droplet size distribution in the emulsion, either for control of creaming rate or for some physical action or characteristic required of the emulsion. This can be accomplished in the homogenizer by passing the product more than once through the valve (Pandolfe and Kinney 1983). Multiple passes do not reduce the modal particle size but reduce the probability of having oversized particles thereby leading to a narrower distribution of sizes. A single pass process minimizes production time and maximizes product consistency but for an extremely uniform particle size as discussed above or for products such as cream liqueurs which require a very small (0.1–0.3 μm) average particle size, it is simply not possible to reach these goals in a single pass through a homogenizer (Masucci 1989b).

If the average particle size is greater than the critical particle size, then the homogenizing pressure should be increased. On the other hand, if there is sizeable tail of particles greater than the critical diameter or a bimodal distribution where the second peak occurs at a particle size in excess of the critical diameter, then the number of passes should be increased (Masucci 1989c). This rule of thumb assumes that the premix quality is good, the formulation is correct and there are no mechanical deficiencies (Masucci 1989c).

In production of stable, high-quality traditional Irish Cream liqueurs by one manufacturer (Fig. 4.1), the optimal homogenization temperature and total pressure and number of passes are considered to be 57 °C, 27.6 MPa (4000 psi) and two, respectively. Whilst the case for two passes is well established, there may well be a case to gain energy savings through the use of a lower homogenization temperature and total pressure of 55 °C and 24.1 MPa (3500 psi), respectively, without detriment to the traditional Irish Cream liqueur stability and quality.

4.3.5   *Troubleshooting Protocol*

A troubleshooting protocol for the production of traditional Irish Cream liqueurs can be specified (Fig. 4.2) based on the scientific and technical literature review and observations at one manufacturer.

4.4   Conclusions

Traditional Irish Cream liqueurs are added value, long-life, oil-in-water emulsions, combining the flavour of an alcoholic drink with the texture of thickened cream. The fundamental studies of a number of workers in the UK and Ireland during the 1980s have enabled the significant commercial problems associated with the production of Irish Cream liqueurs to be overcome. Effective classical high-pressure homogenization and selection of sodium caseinate, either as the sole stabilizer at a protein to fat ratio of *circa* 0.2 or combined with glycerol monostearate at a reduced level, can prevent creaming and fat plug formation. Addition of tri-sodium citrate di-hydrate at 0.19 % by weight in the formulation can prevent emulsion destabilization by calcium-induced aggregation and minimize calcium citrate crystal deposits in an exemplar Irish Cream liqueur. The standard two-stage design "plug-type" homogenizer valve is widely used in the production of Irish Cream liqueurs. For efficient homogenization operation, it is important to achieve a good premix quality, exclude air, control the dispersed phase viscosity and effectively specify homogenization parameters. The use of a high-shear in-line mixer within the process improves premix quality. A homogenization temperature of 55 °C helps to control the dispersed phase viscosity. Effective homogenization results through the use of a total pressure in two stages, typically at 24.1 MPa (3500 psi) and 3.45 MPa (500 psi), and over two homogenization passes. There may be a case for a lower total homogenization pressure (24.1 MPa; 3500 psi) based on energy savings, and the use of a troubleshooting protocol for the production of traditional Irish Cream liqueurs. The commercial importance of cream-based liqueurs will continue to drive scientific and technical developments aimed at lowering the cost of manufacture, reformulating existing products and generating innovative new products.

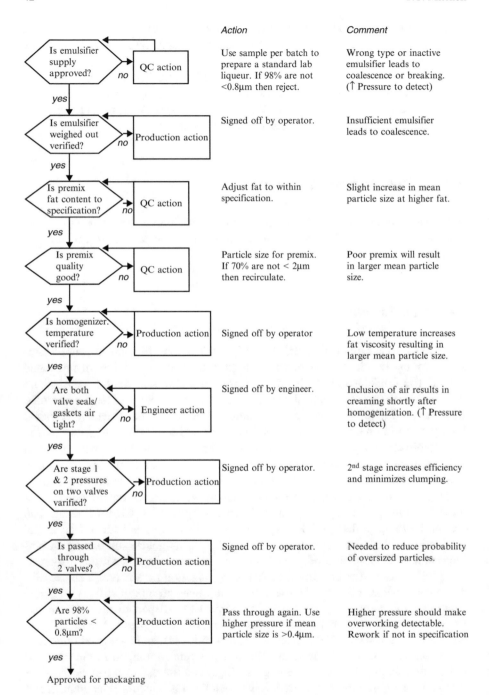

Fig. 4.2 Troubleshooting protocol for the production of traditional Irish Cream liqueurs

References

Alcohol and Tobacco Tax and Trade Bureau (2007) Laws and Regulations under the Federal Alcohol Administration Act. Department of the Treasury, Alcohol and Tobacco Tax and Trade Bureau, Washington, DC, pp 6–167

Anderson M (1991) Functional aspects of milk fat constituents. In: Rajah KK, Burgess KJ (eds) Production, technology and utilisation. The Society of Dairy Technology, Huntingdon, pp 9–17

APV (n.d.a) Effect of air on homogenizing efficiency and product quality, APV, Technical Bulletin, TB-72:1–2

APV (n.d.b) The effect of the second-stage homogenizing valve. APV, Technical Bulletin, TB-58:1–2

APV (n.d.c) An evaluation of emulsifier cost versus processing energy cost. APV, Technical Bulletin, TB-61:1–3

Banks W, Muir DD (1985) Effect of alcohol content on emulsion stability of cream liqueurs. Food Chem 18(2):139–152

Banks W, Muir DD (1988) Stability of alcohol-containing emulsions. In: Stainsby G, Dickinson E (eds) Advances in food emulsions and foams. Elsevier, Barking, pp 257–283

Banks W, Muir DD, Wilson AG (1981a) The formulation of cream-based liqueurs. Milk Ind 83(5):16

Banks W, Muir DD, Wilson AG (1981b) Extension of the shelf life of cream-based liqueurs at high ambient temperatures. J Food Technol 16:587–595

Bergenstahl B (1995) Emulsions. In: Beckett ST (ed) Physico-chemical aspects of food processing. Blackie Academic & Professional, Glasgow, pp 49–64

Clark AH (1992) Gels and gelling. In: Schartzberg HG (ed) Physical chemistry of foods. Marcel Dekker, New York, pp 263–305

Clutton DW (2003) Liqueurs and Speciality Products. In: Lea AGH, Piggott JR (eds) Fermented Beverage Production, 2nd edn. Kluwer Academic/Plenum, New York, pp 309–334

Cunnington J (2010) Comment. A tough future for Baileys and cream-based liqueurs. Euromonitor International, London

Davies DT, White JCD (1958) The relation between the chemical composition and the stability of the caseinate complex. J Dairy Res 25:256–266

Dickinson E (1992) An introduction to food colloids. Oxford University Press, Oxford, pp 2–7

Dickinson E, Stainsby G (1988) Emulsion Stability. In: Dickinson E, Stainsby G (eds) Advances in food emulsions and foams. Elsevier, Barking, pp 1–44

Dickinson E, Narhan SK, Stainsby G (1989) Stability of cream liqueurs containing low-molecular weight surfactants. J Food Sci 54(1):77–81

Early R (1998) Liquid milk and cream. In: Early R (ed) The technology of dairy products, 2nd edn. Blackie Academic & Professional, London, pp 1–49

European Commission (2008) Regulation (EC) No 110/2008 of the European Parliament and of the Council. Official Journal of the European Union, 13 Feb, p.L39/16-L39/54

Euston SR (2008) Emulsifiers in dairy products and dairy substitutes. In: Hasenhuettl GL, Hartel RW (eds) Food emulsifiers and their applications. Springer, New York, p 210

Heffernan SP, Kelly AL, Mulvihill DM (2009) High-pressure-homogenised cream liqueurs: emulsification and stabilization efficiency. J Food Eng 95(3):525–531

Horne DS, Muir DD (1990) Alcohol and heat stability of milk protein. J Dairy Sci 73:3613–3626

Horne DS, Parker TG (1980) The pH sensitivity of the ethanol stability of individual cow milks. Neth Milk Dairy J 71:126–130

Kinsella JE, Whitehead DM (1989) Proteins in whey: chemical, physical and functional properties. In: Kinsella KE (ed) Advances in food nutrition and research, vol 33. Academic, London, pp 343–438

Lynch AG, Mulvihill DM (1997) Effect of sodium caseinate on the stability of cream liqueurs. Int J Dairy Technol 50:1–7

Masucci S (1989a) Importance of premix quality. Homogenizer Digest 5:1–2

Masucci S (1989b) Multiple-pass homogenization. Homogenizer Digest 6:1–5

Masucci S (1989c) Higher pressure or multiple passes. Homogenizer Digest 7:1–5

Medina Torres L, Calderas F, Gallegos-Infante JA, González-Laredo RF, Rocha-Guzmán N (2009) Stability of alcoholic emulsion containing different caseinates as a function of temperature and storage time. Colloids Surf A Physiochem Eng Asp 352:38–46

Muir DD (1989) Cream liqueurs. J Soc Dairy Technol 42(2):31

Muir DD, Banks W (1984) From Atholl Brose to cream liqueurs: development of alcoholic milk drinks stabilised with trisodium caseinate. In: Proceedings of the International Conference on Milk Proteins, Luxemburg, 7–11 May, Centre for Agricultural Publishing and Documentation, Wageningen, pp 120–128

Muir DD, Banks W (1986) Technical note: multiple homogenisation of cream liqueurs. J Food Technol 21:229–232

Muir DD, Dalgleish DG (1987) Differences in behaviour of sodium caseinates in alcoholic media. Milchwissenschaft 42(12):770–772

Mulder H, Walstra P (1974) The milk fat globule. Emulsion science as applied to milk products and comparable foods 1st edition. Commonwealth Agriculture Bureaux, Farnham Royal and Centre for Agricultural Publishing and Documentation, Wageningen

Narhan SK (1987) Instability of dairy emulsions containing alcohol. Ph.D. thesis, University of Leeds, Leeds

Pandolfe WD (1995) Effect of premix condition, surfactant concentration, and oil level on the formation of oil-in-water emulsions by homogenization. J Dispers Sci Technol 16(7):633–650

Pandolfe WD (1999) Homogenizers. In: Francis FJ (ed) Encyclopedia of food science and technology, 2nd edn. Wiley, New York, p 1289

Pandolfe WD, Kinney RR (1983) Recent developments in the understanding of homogenization parameters. Paper presented by Dr. Pandolfe at Summer National meeting, American Institute of Chemical Engineers, Denver, CO, August 23, pp 1–18

Phipps LW (1983) Effect of fat concentration on the homogenization of cream. J Dairy Res 50:91–96

Power PC (1996) The formulation, testing and stability of 16% fat cream liqueurs. Ph.D. thesis (Food Technology), National University of Ireland, Cork

Rothwell J (1989) Cream processing manual, 2nd edn. The Society of Dairy Technology, Huntington, pp 1–141

Rovinsky LA (1994) The analysis and calculation of the efficiency of a homogenizing valve. J Food Eng 23:429–448

Schubert H, Armbruster H (1989) Principles of processing and stability of food emulsions. In: Speiss WEL, Schubert H (eds) Engineering and food, vol 1, Physical properties and process control. Elsevier, London, pp 186–187

Silverson (n.d.) Application report: production of cream liqueurs. Silverson Machines, East Longmeadow. 36FA2:1–4

Towler C (1994) Developments in cream separation and processing. In: Robinson RK (ed) Modern dairy technology, vol 1, 2nd edn, Advances in milk processing. Chapman & Hall, London, pp 61–105

Tunick MH (2009) Dairy innovations of the past 100 years. J Agri Food Chem 57(18):8093–8097

Varnam AH, Sutherland JP (1994a) Beverages: technology, chemistry and microbiology. Chapman & Hall, London, pp 431–433

Varnam AH, Sutherland JP (1994b) Milk and milk products: technology, chemistry and microbiology. Chapman & Hall, London, pp 1–41

Widmar CC, Tripp D (1990) Cream based liqueurs. US Patent Number 4,957,765, 18 Sept 1990, pp 1–5

Wilbey A (1992) Homogenisation. J Soc Dairy Technol 45(2):31–32

Chapter 5
Modernization of Skyr Processing: Icelandic Acid-Curd Soft Cheese

Gudmundur Gudmundsson and Kristberg Kristbergsson

Contents

5.1 Introduction

Skyr is a traditional product which is very popular today but has been a part of the diet in Iceland since the Viking era. It was a staple food in the old agricultural society of Iceland and both sheep and cattle were raised for milk production and the making of skyr (Gisladottir 1999). When the food supply in Iceland was modernized during the last century, skyr proved to be suitable for mass production and distribution.

Skyr is a fresh acid-curd soft cheese made from skim milk. Among related products in other countries are quarg in Germany, tvorog in Russia, and the Arabic labneh (Fox et al. 2000; Nsabimana et al. 2005).

G. Gudmundsson
Department of Food Science and Nutrition, University of Iceland, 101, Reykjavik, Iceland

LYSI hf., Fiskislod 5-9, 101, Reykjavik, Iceland

K. Kristbergsson (✉)
Department of Food Science and Nutrition, University of Iceland, 101, Reykjavik, Iceland
e-mail: kk@hi.is

© Springer Science+Business Media New York 2016
A. McElhatton, M.M. El Idrissi (eds.), *Modernization of Traditional Food Processes and Products*, Integrating Food Science and Engineering Knowledge Into the Food Chain 11, DOI 10.1007/978-1-4899-7671-0_5

Skyr has texture similar to a thick yogurt and is normally consumed as such. This type of cheese is produced by the coagulation of milk proteins by acidification, or the combined action of acid and heat, with or without a small quantity of rennet. Skyr has probably been a part of the Icelandic diet since the first settlers arrived in Iceland more than one thousand years ago (Gisladottir 1999). It is mentioned in medieval Icelandic literature and remnants of products similar to skyr have been found in archaeological excavations of medieval farms in Iceland.

The old society of Iceland was based on agriculture and remained without much change until the end of the nineteenth century. Farms were based on self-sufficiency and most of a farm's produce was consumed on location. In this society, skyr was a staple food and both sheep and cattle were raised for milk production and the making of skyr. Normally at the time only a part of the milk was consumed fresh, and most of it was separated into cream and skim milk, the cream was used for butter, while the skim milk was used to make skyr. In skyr production both curds and skyr whey were collected. The latter was used as a drink or for the preservation of meat by pickling. The importance of skyr whey was probably one of the main reasons why skyr production was considered more economical than production of hard cheeses (Gisladottir 1999).

In the first part of the twentieth century enormous social changes took place in Iceland. Agriculture was modernized, new industries evolved, small towns appeared, and the established ones grew rapidly. This urbanization called for changes in the food production and a new market for food had emerged and this market needed to be supplied. Since cow milk is much better suited for mass production than ewe milk, cow milk became the dominant type of milk produced. At the same time, dairies were established to process and distribute milk to the new market, and gradually took over the production of dairy products from farms.

Although skyr is a traditional product, which was tightly linked to the needs and practices of the old society, it has proved to be suitable for mass production and distribution. Actually it has in the last decade undergone a revival with constantly increasing consumption. In Iceland in 2004 the consumption of skyr and a type of set yogurt called "skyr.is" was 10.7 kg per capita and had then increased more than twofold in ten years (Anon 2005). Still modern technology has shaped this product in such a manner it now suits a completely different world. In 2015 the main dairy in Iceland MS produced 36 million containers of skyr.is where 11.5 million were used for domestic consumption. One needs to view this number with respect to the total population of Iceland which is only 320,000. The rest 24.5 million containers was exported to Europe and USA with about 2/3 to Scandinavia.

5.2   Traditional Production

Up to the beginning of last century two main types of skyr were produced depending on whether rennet was used or not. In "auto-coagulated skyr" rennet was omitted and lactic acid bacteria gradually coagulated the skim milk, but in "coagulated skyr," a small quantity of rennet from calve abomasum's (fourth and final stomach compartment in ruminants) was used to speed up the skyr making. In the production of "auto-coagulated" skyr, a large vessel was used, which took days or weeks to fill with skim milk. Prior to filling of the skyr vessel used for incubation a small amount of fresh skyr was placed at the bottom of the vessel for inoculation and the skyr was allowed to ferment overnight. Sometimes the skim milk was heated to boiling and cooled down before pouring in the vessel, but sometimes heating was omitted. When the vessel was full and the milk had coagulated, the whey was separated from the curds using cloth (Gudmundsson 1914). Production of "auto-coagulated" skyr is not suitable for modernization and this type of skyr probably disappeared at the beginning of last century, when commercial rennet became available and the processing of milk gradually moved from farms to dairies.

Today some traditional production of skyr is still practiced in Iceland, but both the production methods and probably the product itself, differ from that produced by the old production process. Skyr is coagulated with a combined action of lactic acid bacteria and rennet ("coagulated skyr") but in the modern production of skyr.is there is no rennet used. Production on farms these days is almost nonexistent and traditional skyr available at the commercial market is produced in modern dairies, utilizing modern equipment (Fig. 5.1).

The production of traditional skyr may be divided into four steps: pasteurization, incubation, cooling, and cloth filtration (Fig. 5.2). Fermentation and precipitation of curds takes place during the cooling phase. What makes the process and product "traditional" is the use of skyr from an earlier batch as a starter and separation of curds and whey by filtration through cheese cloth. Gudmundsson (1987) describes the traditional production as follows:

> Skim milk was heated to 90-100 °C and then cooled down to 40 °C. Water-diluted skyr from an earlier production was added, approximately 15 gr. per liter of milk along with [commercial] cheese rennet, approximately 6 ml. per 100 liters of skim milk. This was allowed to sour until it reached a pH of about 4.7, taking 4½ to 5½ hours. The liquid was cooled down to 18–20 °C and left for about 18 hours or until the pH reached 4.2. Then the filtering was started by pouring the skyr curd into linen bags and the whey allowed to drain

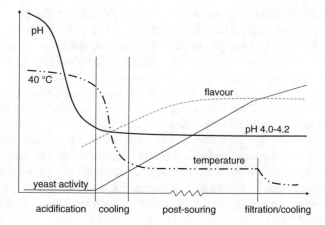

Fig. 5.1 Production of traditional skyr. The figure is for illustrative purposes (Reconstructed from: Magnusson 1986)

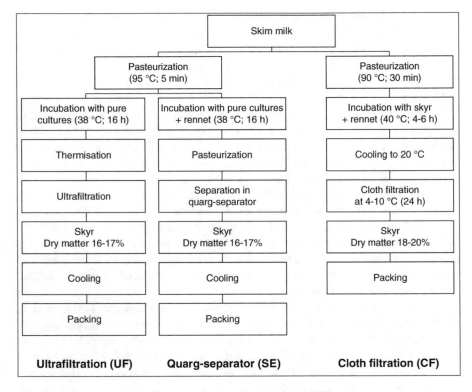

Fig. 5.2 Different methods of skyr preparation (Gudmundsson 2007)

through the bag for approximately 6 hours at a temperature of 19–20 °C, and then for another 18 hours at 6–8 °C.

The total filtering time was about 24 hours, leaving the finished skyr with a pH of 3.8-4.0 and a dry matter of 17–20 %. To make one kilogram of skyr, 5 liters of skim milk were needed.

Considerable variation was and still is between different producers of traditional skyr. Heating and cooling profiles differ, but they are to a certain extent determined by available facilities and equipment. Also quantity and quality of starter and quantity of rennet varies greatly (Petursson 1939a). Petursson (1939a) recommends 0.1–0.15 % starter and 5 ppm rennet and Magnusson (1986) recommends 0.01–0.1 % starter and 50–100 ppm rennet. According to the description above (Gudmundsson 1987) 1.5 % diluted skyr and 60 ppm of rennet would be appropriate, but the dilution factor is lacking, which makes comparison difficult.

The texture of the skyr produced, as described by Gudmundsson (1987), is firm and water or milk has to be mixed with the skyr before serving. The taste is probably much less sour than the taste of the old production styled skyr, but the shelf life is also that much shorter. In the old days it was necessary to keep skyr for months, as the summer production of skyr had to be stored until winter. This skyr which became very sour, called "sour-skyr," was stored in tightly closed barrels, which were sealed with tallow to retard growth of molds (Petursson 1939b; Magnusson 1967a).

5.3   Modern Production

Production of skyr has changed dramatically over the last 80 years. Today the raw material is thoroughly standardized, acidification, heating and cooling accurately controlled and a large part of the production is inoculated with pure bacteria cultures (Fig. 5.2). The main difference between traditional skyr, as described previously, and modern skyr production, in addition to the use of pure starter cultures, lies probably in the separation of curds and whey (Figs. 5.3 and 5.4). The large diaries have modernized this process. This started with the introduction of quarg-separators. As a result the yield increased almost 30 %, although a large part of the whey proteins were still leached out with the whey (Gudmundsson 1987).

Quarg-separators are still in use, but today the bulk of the production is concentrated by ultrafiltration (Fig. 5.2). This method results in a much higher protein yield, since the whey proteins are retained in the curds and loss of casein is minimized. The product is similar to yogurt that is allowed to set in the final retail container. At the same time the protein composition of the skyr has changed because the ultra-filtered skyr (skyr.is) contains a higher ratio of whey proteins than skyr produced by cloth filtration or separation. According to Gudmundsson (2007) the modern skyr.is has a more porous microstructure but similar stability against oscillatory shear. This may be due to the high level of whey proteins.

Fig. 5.3 Tunnel for traditional separation of curd and whey. The skyr curds were placed in linen bags and loaded in the tunnel. The tunnel was rotated in order to speed up separation of curds and whey. The process went on overnight in the skyr separators (Saemundsson 1954)

Fig. 5.4 Ultrafiltration of modern skyr (©MS Selfoss, Iceland)

One large dairy in Iceland uses a method, which is a combination of old and new technologies, where skyr is used as a starter, together with ultrafiltration for the separation of curds and whey.

The modern skyr is without doubt very different from the skyr produced in the old days. The acidification is most likely less and accordingly the taste much less acidic or sour. The texture has also changed, being much softer today. In part, this can be explained by a lower content of dry matter, but the new separation methods also influence the texture making the skyr smoother. Due to the softer texture there is no need for blending skyr with milk or water before consumption. In the last few years a new product "skyr drink," a very thin skyr type consumed as a drink, has become immensely popular especially among school children.

5.4 Research

Through the years relatively little effort has been devoted to scientific research on skyr. In the dairy industry various aspects of skyr production have been studied, but the results have usually not been published in the scientific literature. All the same, skyr research in Iceland has a long tradition. It started 100 years ago with the work of Gudmundsson (1914) on the microbiology of skyr.

Gudmundsson described lactic acid bacteria in skyr, which appeared to be very similar to the principal lactic acid bacteria in Bulgarian yogurt. These were cocci and bacilli, which when used for the acidification of milk gave very comparable results to yogurt bacteria. However, Gudmundsson tried to acidify skyr with pure cultures of these bacteria, but without success. According to his results, unfiltered skyr from pure cultures had a very similar taste to unfiltered ordinary skyr, but after filtration the taste of the test product was inferior to good genuine skyr. Gudmundsson explained this by the lack of yeast activity in the former. He claimed yeast produced ethanol and other volatile chemicals, which improved the flavor of skyr (Gudmundsson 1914). This assertion signified the start of a long debate on the importance of yeast for the quality of skyr.

The most extensive scientific research on skyr took place 25 years later, when Petursson (1938, 1939a, b) studied the microflora of skyr and various aspects of skyr production technology. In accordance with earlier results of Gudmundsson (1914), Petursson found lactic acid bacteria and yeast to be the principal microorganisms in skyr. The lactic acid bacteria were of the types *Streptococcus* and *Thermobacterium*. Petursson found three lactic acid bacteria to be most important. These he classified with almost certainty as *Streptococcus thermophilus*, *Thermobacterium bulgaricum* (*Lactobacillus bulgaricus*), and *Th. jugurt* (*L. jugurt*). Petursson also found many strains of yeast in skyr, but all grew poorly in milk if it did not also contain lactic acid bacteria. Petursson did not consider yeast beneficial for the production or quality of skyr, but rather that yeast could impart off-flavors and limit the shelf life of skyr (Petursson 1938).

Petursson tested single strains and various combinations of bacteria in skyr production. He concluded that the combination of *S. thermophilus* and *Th. jugurt* seemed essential for the production of good quality and high yield, *S. thermophilus* being important for fast fermentation and acidification and *Th. jugurt* for flavor. Adding yeast to the lactic acid bacteria did not improve the product (Petursson 1938).

According to Petursson, pure bacteria cultures could successfully be used to produce skyr. This was tested in a dairy on both pilot and production scales with good results. Further production tests with pure cultures of lactic acid bacteria were probably not carried out. At that time Petursson was already supplying the dairies with pure cultures, which contained yeast besides lactic acid bacteria (Petursson 1938, 1939a).

Petursson (1939a) studied various other factors in the production of skyr, preheating of skim milk, fermentation temperature, use of cheese rennet, and quantity of starter. The main results were as follows:

1. Preheating the skim milk at 90–100 °C for a few minutes gave optimum yield and texture.
2. Fermentation temperature and cooling profile had a major effect on the quality of skyr. Inoculation at a temperature of 40–45 °C with very slow cooling resulted in optimum flavor and texture.
3. At a low fermentation temperature the use of rennet was important for expediting coagulation, but if the milk was set at a high temperature (45 °C), addition of rennet did not appear to be important.

4. Suitable concentration of starter for an inoculation temperature of 40–45 °C was 0.1–0.15 %, based on undiluted skyr. A higher concentration was needed if fermentation temperature was low.

Later Magnusson (1967b) continued with the studies on the microbiology and production of skyr. He isolated bacteria and yeast in skyr and concluded that *S. thermophilus* and *L. bulgaricus* were the most important bacteria in skyr. According to Magnusson the streptococci did not on their own give skyr of good quality. The acidification was normal, but the skyr was dry and lacked flavor. Similarly lactobacilli did not give good skyr. The acidification was too slow, the skyr was too thin and lacked acidity. The combination of *S. thermophilus* and *L. bulgaricus* turned out to be necessary for the production of good quality skyr, although according to Magnusson it lacked the characteristic flavor of yeast. Magnusson tried to answer the question whether yeast was necessary for skyr flavor, but the results were inconclusive. He suggested this depended on whether the skyr was consumed fresh or not. For consumers, which were used to freshly prepared skyr, the yeast flavor was perhaps not a benefit. On the other hand, consumers, which were accustomed to more excessively fermented skyr, might prefer skyr with yeast flavor. Magnusson tried to identify the yeast giving the characteristic skyr flavor and according to his results only *Saccharomyces steineri* gave the typical yeast flavor (Magnusson 1967b).

It is of interest to note, that sixty years after Petursson's research on skyr, the largest dairy in Iceland started producing a type of skyr or set yogurt using pure bacteria cultures without any yeast and no rennet. Today this product line "skyr.is" enjoys great commercial success in Iceland. The elimination of yeast and rennet are probably a part of an effort to optimize shelf life and minimize quality fluctuations. Yeast activity would probably compromise both and rennin activity might limit shelf life by imparting a bitter off-flavor to the product.

5.5   Regulations and Composition

Icelandic regulation on milk and dairy products include clauses on skyr criteria. According to these, skyr has to be produced from pasteurized skim milk and acidified with skyr starter. The use of cheese rennet is permitted. The minimum content of milk solids is specified as 16 % (Ministry of Industries and Innovation 2012).

Unblended skyr contains typically 80.1 g water, 14.6 g protein, 4.3 g carbohydrates, 0.8 g ash, and 0.2 g fat per 100 g. The ready-made variety has a lower content of dry matter, and typically contains 83.3 g water, 11.5 g protein, 3.3 g carbohydrates, 0.8 g ash, and 0.2 g fat per 100 g (Reykdal 2003).

Skyr contains significant levels of some vitamins and minerals. In 100 g of unblended skyr there are approximately 0.10 mg thiamine, 0.29 mg riboflavin, 100 mg calcium, 190 mg phosphorous, 150 mg potassium, and 0.4 mg zinc (Reykdal 2003).

References

Anon (2005) Report on consumption of dairy products. Association of the Icelandic Dairy Industry

Fox PF, Guinee TP, Cogan TM, McSweeney PLH (2000) Fundamentals of cheese science. Aspen, Cork

Gisladottir H (1999) Icelandic food tradition (in Icelandic). Mal og menning, Reykjavik

Gudmundsson G (1914) Icelandic and foreign skyr (in Icelandic). Bunadarrit 28:1–16

Gudmundsson B (1987) Skyr. Scandinavian Dairy Ind 4(87):240–242

Gudmundsson G (2007) Rheology and microstructure of Skyr. M.S. Thesis. Department of Food Science and Human Nutrition, University of Iceland

Magnusson S (1967a) Skyr in the past and in the future (in Icelandic). Arbok landbunadarins 1967(1): 69–77

Magnusson S (1967b) Skyr and quality factors in skyr production (in Norwegian). Thesis. Milk Department, Agricultural University of Norway

Magnusson S (1986) On the production of skyr (in Icelandic). Mjolkurmal 1986(2):13–18

Ministry of Industries and Innovation (2012) http://www.reglugerd.is/reglugerdir/allar/nr/851-2012. Accessed 20 November 2015.

Nsabimana C, Jiang B, Kossah R (2005) Manufacturing, properties and shelf life of labneh: a review. Int J Dairy Technol 58(3):129–137

Petursson SH (1938) Milk research and milk microbiology (in Icelandic). In: Report of Industry Department. University of Iceland, Reykjavik, pp 48–54

Petursson SH (1939a) Milk research and milk microbiology (in Icelandic). In: Report of Industry Department. University of Iceland, Reykjavik, pp 49–63

Petursson SH (1939b) Milk theory (in Icelandic) in Mjolkursolunefnd. Isafold. Reykjavik. pp 142–153

Reykdal O (2003) The Icelandic Food Composition Database (ISGEM). MATIS — Food Research, Innovation & Safety, Reykjavik

Saemundsson J (1954) Cheese and Skyr. In: Tomasson H, Saemundsson J, Vestdal JE, Gudjonsson SV (eds) No food is better than milk. (In Icelandic). Framleiðsluráð Landbúnaðarins, Reykjavik, p 40

Chapter 6
Karachay Ayran: From Domestic Technology to Industrial Production

I.K. Kulikova, S.E. Vinogradskaya, O.I. Oleshkevich, L.R. Alieva, and Anna McElhatton

Contents

6.1   The History of Karachay Ayran

Karachay Ayran is a traditional fermented milk product that is popular and widely available within the territory of the Russian Karachay-Cherkess Republic (the North Caucasus region). Ayran microflora are known to consist of a symbiotic mixture of lactic acid bacteria and yeasts. The main features of Karachay Ayran are its characteristic taste and beneficial dietary properties.

There are several theories on how Ayran has appeared in the North Caucasus region. One suggestion is that Ayran was brought by Tartars, while another indicates that it was brought by Turks from Middle East. But the majority of ethnographers consider Ayran to be an indigenous product of Karachai and Balkar people (Tekyev 1989). Traditionally Ayran production was associated with cattle breeding in the region, in that a portion of the milk produced was transformed into this product. In 1907 Kara-Vasilyev who was a veterinary surgeon and researcher of Karachay

I.K. Kulikova (✉) • S.E. Vinogradskaya • O.I. Oleshkevich • L.R. Alieva
FSAEI HPE, North-Caucasus Federal University, Stavropol, Russia
e-mail: kik-st@yandex.ru

A. McElhatton
Faculty of Health Sciences, University of Malta, Msida, Malta

© Springer Science+Business Media New York 2016   55
A. McElhatton, M.M. El Idrissi (eds.), *Modernization of Traditional Food
Processes and Products*, Integrating Food Science and Engineering Knowledge
Into the Food Chain 11, DOI 10.1007/978-1-4899-7671-0_6

mountain dwellers culture, wrote: "A lot of families eat only Ayran. Beverage Ayran ... has an exceptional nutritional value and medical properties." He also noticed that the peoples of Karachay-Cherkess region used Ayran for treatment of diarrheal disease, burns, venomous snake bites and insects stings, severe intoxication, etc. (Kara Vasiliev 1907).

Modern researchers have shown that the symbiosis of Ayran microorganisms has the capacity to inhibit the growth of opportunistic and pathogenic microorganisms (Vinogradskaya et al. 2002). Moreover antimicrobial properties of Ayran are said to be comparable with the antimicrobial properties of acidophilic milk (*Lb. Acidophilus*) and kefir (Table 6.1). The activity of Ayran microflora against spoilage microflora has also been reported. It is thought that this characteristic is one of the reasons why Aryan does not spoil during long-term storage.

In addition, recent literature (Grishina et al. 2011) has also suggested that Ayran may exhibit anticancer effects. The authors reported that the liquid whey component derived from Ayran was effective in reducing the DNA damage when applied to an in vitro colon cell culture model.

6.2   Ayran Home Production

There are several kinds of Aryan made by the Karachays with each variety having a distinct name: Dzhuyrt, Susap, Tuzluk, and others (Fig. 6.1). Dzhuyrt is a very dense product. It is made by carefully separating the whey, formed after fermentation and retaining the curd Susap has a semi-liquid consistence as the curd and whey are mixed; there are instances where water is added to get the right consistency. Susap is a liquid and considered to be a very good refreshing beverage. Tuzluk is a salted Ayran which has an extended shelf life. This form of Ayran is salted, boiled and cooled and can be kept at room temperature for several months. Dzhuyrt is considered to be the best of all Ayran types that can be used as a starter. One or two spoons of Dzhuyrt added to the cooled boiled milk and the milk and mixed thoroughly is sufficient to get the cheese making process started.

Table 6.1 Diameter of zones of inhibition (mm) produced by microflora of fermented milk products on the test strains as assessed by the disc diffusion method[a]

	Diameter of zones of growth inhibition (mm)		
Fermented milk product	*Escherichia coli*	*Staphylococcus aureus* subsp. *aureus*	*Bacillus megaterium*
Acidophilic milk	16±1.0	17±1.2	17±1.0
Kefir	13±0.6	12±0.5	17±1.5
Homemade Ayran	14±0.5	16±0.8	21±2.0
Homemade Ayran (after deacidification)	13±0.3	14±0.5	17±0.4
Ayran filtrate (after boiling)	14±0.6	16±0.5	20±2.0
Control	8±0.5	8±0.5	8±0.5

[a]Reconstructed from Vinogradskaya et al. (2002)

Fig. 6.1 Different types of
Karachay Ayran

   The container with the milk is covered with a woolen blanket to keep it warm
and the milk gradually cools, allowing all groups of microorganisms to grow. The
process takes about 24 h. The taste of domestically produced Ayran varies as a lot
of factors influence Ayran properties: type of milk, milk temperature, speed of
cooling, and other. Such technology is acceptable for homemade beverages, but
not for commercial products.

6.3   Adaptation of the Homemade Karachay Ayran Technology to Commercial Conditions

Commercial Ayran is a fermented milk product made by a mixed (lactic acid and
yeast) fermentation using pure cultures of lactic bacteria (*Streptococcus salivar-
ius* ssp. *thermophilus, Lactobacillus delbrueckii* ssp. *bulgaricus*) and yeast
(Russia. NSso: GOST R 53668-2009 Ayran specifications. State Standard of
Russian Fed.). The product can be salted or mixed with mineral or drinking water.
Its characteristics are determined by the Russian Federation legislation and technical documentation.

   The main stages of Ayran production are similar to the production of other
fermented milks (Fig. 6.2). These technologies commonly used have been widely
described by many authors (Wszolek et al. 2006; Hutkins 2006). The specificity of

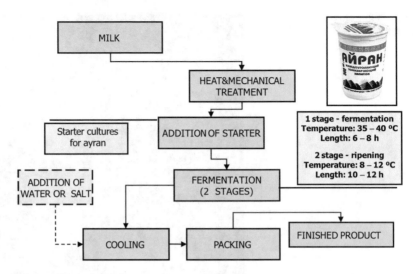

Fig 6.2 The commercial Ayran technology

Ayran technology lies in its two-stage fermentation: cultivation of thermophilic bacteria at 35–45 °C and ripening at 8 °C for yeast growth. But the taste of a commercial
product is said to differ significantly from that of the homemade Karachay Ayran.

To adapt the homemade Karachay Ayran technology to commercial conditions
and yet retain the flavor associated with domestically produced Ayran two main
issues had to be solved. They are the identification of the real Karachay Ayran
microflora and the adjustment of technological parameters to produce the characteristic and accepted taste of the domestic product.

Research based on the analysis of 37 homemade Ayran samples collected in various places of Karachay-Cherkess Republic (the North Caucasus region) showed
that most of the samples had different organoleptic properties and microflora with
various morphological characteristics. Some samples had a thick consistency; other
samples were characterized by the presence of highly noticeable yeast flavor and a
slightly frothy consistency. But nevertheless all samples had a specific flavor which
differed from the taste of other well-known fermented milk products. The then
general features of typical Ayran microflora composition were identified as follows:
The major morphological types of microorganisms were found to consist of streptococci, diplococci, rod bacteria, and yeasts. As regards composition, the bacilli
(rods) and yeasts were reported to be the dominant microorganism present and were
present well in excess of any other cocci present. Of the morphological groups of
microorganisms observed, significant diversity was noted, especially among bacilli.
Ripened Ayran when compared with kefir was found to have comparable population
densities of bacteria and yeasts.

More specifically, Karachay-Cherkess Ayran microflora consisted mainly of
Lactococcus lactis subsp. *lactis*, *Lactococcus lactis* subsp. *lactis* biovar. *diacetilactis*,
Lactobacillus delbrueckii subsp. *Bulgaricus*, Streptococcus thermophilus, and

non-lactose-fermenting yeast. The major groups of microorganisms were visualized in the images obtained by electron microscopy (Fig. 6.3, a, b). Thus it can be said that Ayran is produced through the symbiotic relationship between thermophilic and mesophilic lactic acid bacteria and yeasts (Vinogradskaya et al. 2002).

The use of three temperature intervals corresponding to the temperature optimums of each microorganism type (Fig. 6.4) was proposed to create a commercial technology to mimic that of domestically produced Ayran.

The high temperature at the first stage would contribute to the formation of the thick consistency. At the second stage mesophilic microflora would create the characteristic beverage taste, with the final stage organoleptic characteristics completely formed by the yeasts.

The temperature of first two stages would need to be optimized to obtain a product comparable to the homemade Ayran. Skim milk was fermented with a starter comprising microorganisms of Karachay Ayran. The milk was pasteurized, and then

Fig. 6.3 (a, b) Microflora of Karachay Ayran

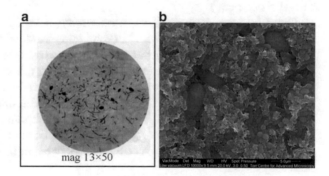

mag 13×50

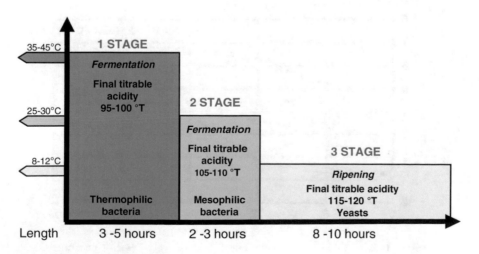

Fig. 6.4 Three stages fermentation technology

cooled to the required fermentation temperatures. Then the 3 mL of a starter was added to every 100 mL of milk. The fermentation temperatures at the first and second stages varied within the range 35–45 °C and 25–30 °C, respectively. In all the cases ripening temperature (the third stage) was 8–10 °C (according to commercial Ayran technology). Fermentation was stopped when titratable acidity of milk curd reached 95–100 °T at the first stage; 105–110 °T at the second stage; and 115–120 °T at the third stage. Product sensory evaluation was used as an output parameter.

The results showed that if the temperature of first stage was more than 40 °C, the product had an undesirable consistency with significant syneresis. When the second stage temperature was less than 30 °C the speed of acidification was higher but the taste was less intense. The analysis of data produced would indicate that (38 ± 2) °C is the optimum temperature at the first stage, and (25 ± 2) °C—at the second stage.

At the optimal temperature conditions the ratio of the major microorganism groups during fermentation, ripening, and storage were close to homemade Ayran (Fig. 6.5). *Lactobacillus delbrueckii* subsp. *bulgaricus* plays a leading part promoting a rapid milk clot formation within 3.5–4.5 h and dominated in the finished product $(10^7 – 10^8$ per 1 cm^3).

Cocci forms of bacteria (*Lactococcus lactis* subsp. *Lactis, Lactococcus lactis* subsp. *Lactis* biovar. *Diacetilactis*) developed more slowly; in the finished product their number didn't exceed 10^4 per 1 cm^3. Yeast, as expected, were accumulated and activated during product ripening and storage. On the second day of storage the yeast and coccus amounts were nearly equal (Vinogradskaya et al. 2002).

Fresh Ayran has a pleasant sour milk flavor, and a thick consistency with gas bubbles. During storage the yeasts become the predominant microorganisms so the product acquires a light yeasty taste and specific odor. Thus, the recommended temperature conditions allow to produce a fermented milk with typical properties of Karachay Ayran.

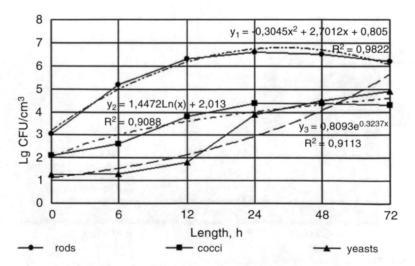

Fig. 6.5 Changes of the main microorganism groups during Ayran fermentation and ripening

Though commercial Ayran is available, the discerning consumer these days tends to prefer authentic tastes. For this reason the modernization and commercialization of what otherwise is a domestically produced dairy product has to be approached with care. Effort is being made to recapture that which was commonly available to rural and nomadic communities of the past, will with modernization be available to the twenty-first century consumer who could benefit from the nutrition and other benefits associated with this product.

References

Grishina A, Kulikova I, Alieva L, Dodson A, Rowland I, Jin J (2011) Antigenotoxic effect of Kefir and Ayran supernatants on fecal water-induced DNA damage in human colon cells. J Nutr Cancer 63(1):73–79

Hutkins RW (2006) Cultured dairy product. In: Hutkins RW (ed) Microbiology and technology of fermented foods. Blackwell, Oxford, pp 107–144

Kara Vasiliev I (1907) Karachay Ayran. Herald of Public Veterinary Medicine 16:564

Russia. NSso: GOST R 53668-2009 Ayran specifications. State Standard of Russian Federation

Tekyev KM (1989) Karachay and Balkar. Traditional life-support System. Nauka, Moscow, pp 260–285

Vinogradskaya SE, Evdokimov IA, Evsukova AN, Levchenko JV, Gorbacheva ES (2002) Homemade Ayran Microflora. Proceedings of the 2 All-Russian Scientific-Practical Conference Modern Advancement of Biotechnology. NCSTU, Stavropol, Russia, 2, pp 89–92

Wszolek M, Kupiec-Teahan B, Guldager HS, Tamime AY (2006) Production of Kefir, Koumiss and other related products. In: Tamime AY (ed) Fermented milks. Blackwell, Oxford, pp 174–216

Chapter 7
German Bread and Related Process Technology

Karl Georg Busch

Contents

7.1   Introduction

Up until around 250 years ago, bread was the most important staple food in Germany with its importance transmitted through traditional fairy tales and children's songs for centuries (Goetz 1973). In rural areas, bread used to be baked in stone ovens, which were used by all the village inhabitants, until around 150 years ago. The ovens were heated to the required temperature by means of a wood fire, after which the ash was removed and the bread subsequently baked in the hot oven. A lack of bread meant that people went hungry (Jacob 1954).

With the increase in the number of settlements and the growth of towns in medieval times, the sale of bread became more and more important as many families no longer had the possibility of baking their own. Similar to other craftsmen in that era, bakers formed guilds which ensured that the production of bread was controlled and organized to a certain extent. Although bread was initially only produced by craft

K.G. Busch (✉)
Beuth University of Applied Sciences, Berlin, Germany
e-mail: kbusch@beuth-hochschule.de

© Springer Science+Business Media New York 2016
A. McElhatton, M.M. El Idrissi (eds.), *Modernization of Traditional Food
Processes and Products*, Integrating Food Science and Engineering Knowledge
Into the Food Chain 11, DOI 10.1007/978-1-4899-7671-0_7

bakers, the technical innovations during the nineteenth and twentieth centuries in particular led to the baking process being simplified and improved. The milling of cereals and the production of bread and other baked goods by craft bakers had such importance that Museum in Ulm has been dedicated to German Bread.

Nowadays in Germany, baked goods are produced not only in craft or medium-sized bakeries but also in industrial bakeries. As the size of bakeries increases, the process is increasingly mechanized from the production of dough to the bread itself. While the variety of bread produced in craft bakeries on a daily basis is often quite large and only a small number of loaves of each type of bread are made, industrial bakeries produce a small range of bread but bake tens of thousands of loaves of each type. In order to ensure that daily production is maintained at a consistently high standard, the appropriate logistical and constructional facilities are required to deal with the quantities involved. In Germany, 94 % of the population eat bread once or several times a day (Source: *GMF-Mehlreport 09*) while 72 % also eat bread rolls, pastries, and cookies (Source: *GMF-Mehlreport 11*). The daily quantities of bread consumed in 2008 were 178 g for men (3–4 50 g slices) and 133 g for women (2–3 slices), which translates into an annual consumption of 65.0 kg for men and 49.8 kg for women (Source: *Nationale Verzehrsstudie Max-Rubner-Institut Bundesforschungsinstitut für Ernährung und Lebensmittel* (2008)).

A prerequisite for good-quality bread are raw materials of high quality, correct storage of the raw materials up to the time when they are processed, the exact metering of the ingredients, the kneading and proofing processes, and the conditions during baking. In Germany, bread is traditionally made almost entirely from baker's wheat and rye, with spelt also being used in Southern Germany. Kamut, einkorn, and emmer (also known as farro) are varieties of wheat that are becoming increasingly important in the production of "organic" bread and other bakery goods. In the 2008/2009 cereals business year, 6178 million metric tons of soft wheat and 0.884 million metric tons of rye were milled for bread production in Germany (Source: *Verband Deutscher Mühlen* 2009). The much higher proportion of wheat being milled illustrates the overriding importance of wheat as a raw material for bread production.

Commercially produced flour must comply with the grades used to identify the ash content which are laid down in legal regulations. Around 85 % of the wheat flours used for baking are white flours with a low ash (mineral) content with which well-leavened, light bread can be made. Owing to the morphological structure and color of rye grain, rye flours always have a higher ash content and are darker in color than wheat flours. Thus, rye bread is darker than wheat bread.

Five different grades of wheat and rye flour are on the market as well as coarse meal (without the germ) and whole wheat flour. The great variety of grades and blends of flour has resulted in more than 200 types of bread being produced in Germany (Täufel et al. 1993a, b). Bread made from rye flour is more frequently produced and eaten in Northern Germany while bread made from white wheat flours is common in Southern Germany, with only small quantities of rye flour being used.

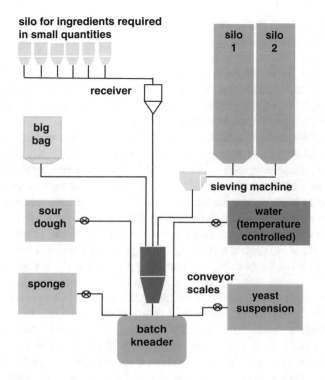

silo for ingredients required in small quantities

silo 1   silo 2

receiver

big bag

sieving machine

sour dough

water (temperature controlled)

sponge

conveyor scales

yeast suspension

batch kneader

Fig. 7.1 Example of a metering arrangement for major and minor ingredients (Busch 2010)

7.2   Dough Production

The production of dough for bread begins by metering the individual ingredients such as flour, water, yeast, and salt; a leavening agent or special ingredients such as oil seeds or other types of grain are also frequently added (Fig. 7.1).

In craft bakeries, the individual ingredients are frequently weighed entirely by hand, with liquid ingredients being measured volumetrically in graduated containers. The dough strength is controlled by means of the quantity of water added and is checked manually. In many bakeries, the dough strength can be corrected by the subsequent addition of water or flour. If the dough is also being processed manually, any deviations from the desired dough strength can be remedied as required.

The uniformity of the dough characteristics is of great importance in larger bakeries where machines are used to make and process the dough. It requires a great deal of effort to compensate mechanically for any variations in dough viscosity or temperature. Hence, the ingredients need to be metered and temperature controlled with a great deal of precision. Most factories employ computer-controlled metering systems which ensure that the same quantities are always used, assuming the system works correctly, and the same dough composition and properties are achieved. There are two different methods of metering: either volumetric metering or gravimetric metering.

The method chosen depends on the dough consistency (either granular, paste-like or liquid) and the metering quantity. The volumetric method is primarily used to meter liquid/paste-like ingredients while granular additions can be metered either volumetrically or gravimetrically (Busch 2010).

The kneading process is divided into three phases: mixing, swelling, and kneading, respectively. During the mixing phase, the ingredients are uniformly distributed. The surface of the flour components is moistened, low-molecular substances dissolve, and the metabolic activity of the yeast increases. The moistening and limited swelling of the protein in wheat doughs leads to the development of a three-dimensional elastic–plastic network. The protein in wheat flour (gluten) governs the processing properties of the dough and to a large extent determines the quality of the bread. The increase in the elasticity of the protein caused by swelling marks the end of the mixing phase and the transition to the swelling and kneading phases. During kneading, the kneading elements generate shear stress fields which not only cause the dough to be torn apart and compressed but also create laminar movement of dough layers relative to each other. At the same time, mechanical energy is converted into thermal energy so that the dough temperature rises. The temperature of wheat doughs should be between 26 and 31 °C. Dough temperatures below 26 °C result in a reduction in the proofing performance of the yeast and thus in longer proofing times. Dough temperatures over 31 °C promote yeast performance and enzyme activity which can impair the quality of the dough (Belitz und Grosch 1987).

In rye doughs, the swelling of the proteins does not lead to the development of a viscoelastic matrix. Instead, malleable doughs, frequently with a moist surface, are formed thanks to the high water absorption of the pentosans. Owing to the lack of gluten in rye flour, energy of the order of 5 Watt Hours per kilogram (Wh/kg) dough is required to produce a rye dough, while between 11 and 15 Wh/kg dough is required for wheat doughs.

The kneading times depend on the geometry and speed of the kneading tool. A distinction is made between kneaders with a single kneading tool (e.g., single-arm kneaders) and kneaders with two kneading tools (e.g., spiral kneaders) or mixers.

7.3   Kneading and Production

The time taken to knead the dough is reduced if the speed is increased. Wheat doughs are produced in intensive kneaders operating at speeds of up to 3000 rpm. The kneading time is limited to 120–180 s. In high-speed kneaders operating at speeds of 120–250 rpm, the kneading time for wheat doughs increases to around 4–6 min. Rye doughs are produced in low-speed kneaders as excessively high shear forces can result in soft and sticky doughs. In addition to discontinuous kneaders, continuously operating systems have been available for several years and are used where production times are sufficiently long and there is no change in the recipe. Systems such as these (known as "continuous kneaders") may comprise a mixer and the actual kneader. The ingredients are conveyed continuously through metering units to the mixing section where they are blended, after which they are conveyed to the kneader in which energy is used to turn them into dough. The dough leaves the kneader in an endless strand. The main advantages of continuous kneading are primarily the fully automatic mode of operation, uniform and hygienic production, and the savings in time and manpower (Klingler 2010).

After production, the dough is left to stand for 5–10 min when it absorbs any water that had not been completely bound so that its surface becomes drier and easier to handle. In wheat doughs, the cross-linking of the gluten increases even further, enhancing the volumetric expansion of both the dough and the bread.

The next stage of production involves portioning the dough. This is still done manually in some craft bakeries but otherwise dough is portioned by means of volumetric dough dividers. The dough is drawn or conveyed from a funnel into a chamber, the volume of which is adjusted in accordance with the desired weight of the dough and taking account of its density. The loss in volume caused by the evaporation of water from the bread during baking must be taken into account during portioning. The dough portions are subsequently formed step by step manually or by machine to obtain the desired dough shape and the shape of the bread required after baking. To this end, wheat doughs are rounded to form balls of dough and the tension in the gluten increases as a result of being stretched and elongated. At the same time, the fermentation gases are partially driven out of the dough and large gas bubbles collapse to form smaller ones. The pore structure of the crumb thus becomes finer and more uniform. Rounding can either be carried out manually or mechanically by conical dough rounders or, in the case of soft wheat doughs and rye doughs, by belt rounders. Both methods have been used successfully in industrial bakeries. Conical dough rounders comprise a grooved horizontal cone around which a guiding plate runs in a spiral, from the largest cone radius to a smaller one. The rotation of the cone causes the dough piece to be conveyed upwards along the guiding plate from the largest cone radius. The dough is mechanically manipulated and formed into balls. Belt rounders have two belts which are arranged at an approximately 90° angle to each other and run in opposite directions at different speeds. The direction in which the dough is conveyed depends on the direction in which the faster belt is moving. The dough is mechanically manipulated by the belts moving in opposite directions and formed into balls.

In order to obtain bread with an elongated shape and a fine and uniform pore structure, the rounded dough is first rolled out between two counter-rotating rollers and then rolled up in a long molder. Long molders comprise a curling net which rolls up the rolled-out dough passing under it along a belt. A second belt running in the opposite direction and at a slower speed above the conveyor belt can be used instead of a net. The distance between the belts decreases in the direction in which the dough is moving so that the laminated dough is rolled up by the upper belt.

Fermentation gas is expelled from the dough as it is being rolled out and subsequently rolled up, thus reducing the size of the pores in the dough (Freund 1995).

Another method consists of cutting a laminated length of dough into portions. Each dough piece is then individually rolled up by a long molder (Klingler 2010). The resulting dough pieces can be proofed either individually, in pans or placed such that the sides of the dough pieces are touching. Rye doughs are worked less intensively to shape them as they do not contain any elastic gluten.

The pore structure defining the texture of the baked product is formed when the dough pieces are proofed. The gases present in the dough after rounding and long molding and the carbon dioxide that develops when the yeast ferments are more or less retained in the dough. In wheat doughs, the gas is retained by a thin film of protein (gluten film) on the inner surface of each pore. The volume of the dough, and thus of the bread, is inversely proportional to the gas permeability of the protein film. In rye doughs, it is the high viscosity of the dough that enables the gas to be retained. Rye doughs, and thus rye bread, are always less well leavened than wheat doughs and wheat bread. The shaped dough pieces are placed in proofing cabinets or conveyed continuously through proofing chambers for the proofing period. The parameters of relevance for proofing are the air temperature (30–40 °C) and a high relative humidity (r.h. approx. 80–90 %). Higher temperatures reduce the proofing times. If the humidity is too low, the surface of the dough can dry out and split, but if it is too high droplets can form on the surface of the dough as a result of condensation and the dough pieces may stick to the proofing trays. The air velocity in the proofing chamber should be between 0.5 and 1 m/s. The proofing times for bread vary between 30 and 60 min but may be longer in some cases (Klingler 2010).

In order to add flexibility to the production of baked goods, the optimum proofing time can be postponed by storing the prepared dough at temperatures between +8 and −5 °C for up to 48 h. The lower temperatures, which must be reached in the center of the dough pieces, inhibit the metabolic activity of the yeasts and enzymes in the dough. During storage, the dough must be prevented from drying out and shrinking by adjusting the air velocity and humidity as required. This procedure is known as retarding and is primarily used when making wheat rolls and pastries.

Interrupting the proofing process enables the dough pieces to be stored for longer periods of time. Temperatures of between −10 and −20 °C are reached in the dough so that the yeast and enzymes are completely inactivated. After freezing, the dough pieces can be stored for several months (Klingler 2010). The bread doughs used for this procedure are frequently prebaked, i.e., the dough is baked for around two-thirds of the required baking time prior to freezing.

The baking process uses heat to transform the shaped and leavened dough pieces into an easily digested and tasty product that will keep. The necessary physical, chemical, biochemical, colloidal, and microbiological processes are initiated in the dough pieces when the temperature rises due to the heat input and are controlled by the baking regime. The physical conditions in the oven are characterized by air temperature, humidity, air velocity, and the temperature of the upper and lower surfaces of the baking chamber (Tscheuschner 1996).

Table 7.1 Processes in dough during baking

Temperature range	Processes in dough during baking
Up to approx. 50 °C	Swelling and dissolution processes
	Enzyme reactions, yeast fermentation
50–90 °C	Yeasts and enzymes are inactivated
	Protein is denatured
	Starch is gelatinized
Around 100 °C	Water evaporates
	The crust sets
>110 °C	Maillard reactions, dextrinization, caramelization

Source: *Grundlagen der Getreidetechnologie*, R.W. Klingler (2010)

Baking is the process of transferring thermal energy to the dough (Table 7.1). The energy used can be divided into required energy and variable energy. The required energy is the amount of energy needed to gelatinize the starch, denature the protein, and partially evaporate the water in the dough. The variable energy comprises the energy used to generate the water vapor, for the air exchange between the baking chamber and its surroundings, to heat the oven, the radiation of heat and the temperature of the exhaust gases. The sum of the required energy is, for example, 178 Wh/kg for rye-based bread and 142 Wh/kg for whole wheat bread baked in pans. The variable energy is frequently two to three times greater than the required energy (Klingler 2010). Thermal energy in ovens is transferred by thermal radiation, conduction, convection, and condensation. Depending on the design of the oven, the primary heat source is either radiation or convection. As the dough absorbs thermal energy, the temperature on the surface of the dough rises and a temperature gradient develops between the crust and the crumb. The temperature of the crust can reach around 180 °C while the temperature of the crumb does not exceed 100 °C. The increase in the dough temperature at the beginning of the baking process causes the yeast to produce more carbon dioxide which, together with the gas already present in the dough, expands as the temperature rises. The volume of the dough therefore increases during the first phase of the baking process. The high temperature causes the surface of the dough to dry rapidly, as a result of which its mechanical stability increases and flavor and color compounds are formed due to the Maillard reaction. The transmission of heat into the center of the dough is accompanied by mass transport. In addition to the conduction of heat through the pore walls, energy is transported into the center of the dough by water evaporating and condensing in each pore. Swelling and dissolution processes occur up to dough temperatures of around 50 °C at which there is also an increase in the activity of hydrolytic enzymes and in yeast fermentation. The fungi (yeasts), microorganisms (lactic acid bacteria), and enzymes ($\alpha + \beta$-amylases, proteinases, polyphenoloxidases) are inactivated at temperatures between 50 and 90 °C. The protein is denatured and the water expelled from the dough as a result is absorbed by the starch in the gelatinization process. The crumb of the bread develops from the dough. Temperatures around 100 °C cause the water to evaporate from the dough throughout the baking time, resulting in a loss of volume. On the surface of the bread, the water content of the crust is low enough to enable temperatures far exceeding 100 °C to be reached. In addition to the Maillard reactions, dextrinization (thermal degradation) and caramelization of the starch occur (Klingler 2010; Ternes 1990).

Ovens used for baking differ in design, type of heat transfer, and operation. In craft bakeries, the most commonly used types are multi-deck ovens, reverse ovens, and rotary ovens. Multi-deck ovens comprise two or more baking chambers arranged one above the other in which the upper and lower heat can be controlled separately. Some ovens are also equipped with fans which can be switched on in order to shorten the baking time. This type of oven can only be loaded from the front. In the case of reverse ovens, the trays can be removed at the front of the oven for loading and, in some models, at the back of the oven for emptying. Rotary ovens bake very evenly and quickly thanks to the high level of air circulation (up to 3000 m^3/h) due to convection. Continuous ovens with lengths between 12 and 50 m are used in industrial bakeries. The width-to-length ratio is generally not less than 1:10. Continuous ovens are usually divided into several zones in which the baking temperature and humidity are set individually. The dough/bread is conveyed from the loading point through the different temperature zones to the unloading point on a continuous net belt. The temperature at the beginning of the baking process is generally around 240–280 °C higher than at the end, when it drops to 220–200 °C. Some types of bread are pre-baked in a preliminary oven before being placed in the continuous oven. The temperature in the preliminary oven can be set at between 300 and 450 °C so that a crust is formed in as little as 2–3 min which also promotes the development of flavor compounds. Baking times range from 15 to 25 min for wheat rolls and pastries, increasing in line with the size of the baked product and the proportion of rye. Rye-based bread and rye bread are both baked for between 50 and 90 min.

Irrespective of the type of oven being used, water vapor is fed into the baking chamber at the beginning of the baking process. Condensation (approx. 10 g/kg bread) occurs when the temperature on the surface of the dough drops below the dew point, thereby releasing the heat of condensation and raising the temperature of the surface of the dough. As a result, the starch is gelatinized, the protein denatured and the surface of the dough becomes elastic and mechanically more stable. The starch undergoes thermal dextrinization and the low-molecular dough components dissolve in the condensate, producing a smooth bread surface with a glazed appearance after the condensate has evaporated (Klingler 2010; Tscheuschner 1996).

In Germany, typical baked goods are leavened rye bread, whole meal bread, pumpernickel and lye dough products. The procedure for making dough and bread from rye flour differs from that used for wheat flour as each type of flour has a different composition and different properties. In the case of wheat flour, the dough properties are primarily determined by the quantity and quality of the gluten and only to a lesser extent by swelling materials. Rye does not develop gluten but has a higher proportion of swelling materials which have pronounced water-binding properties. It is the swelling materials in rye flour that are responsible for the development of the dough. Rye doughs are far more malleable than wheat doughs. The starch in rye flour gelatinizes at temperatures between 55 and 60 °C which means that more starch can be hydrolyzed by α-amylase activity. The crumb of the bread may therefore tear, resulting in cavities.

Lowering the pH in the dough to less than pH 5 reduces the amylase activity. A fermented starter, known as sourdough, is therefore traditionally added to rye doughs. Homo- and heterofermentative lactic acid bacteria, which produce lactic acid, acetic acid, carbon dioxide, and other metabolites, cause fermentation of the rye flour and provide the flavor or flavor precursors that give leavened rye bread its typical flavor. For centuries, the starter has been made in three stages. A basic starter is made with an inoculum consisting of lactic acid bacteria. Flour and water are added after a few hours to make a larger piece of dough that ferments overnight. Prior to making the final dough, the fermented dough is mixed with rye flour and water once more, kneaded, and left to ferment for 3–4 h. This process allows formation of the acid needed to lower the pH in the bread dough. It has been demonstrated that it is best to incorporate around 45–50 % of the flour needed to make sourdough for bread dough in three stages. In order to rationalize the production of sourdough, two-stage and one-stage methods have been developed as substitutes for the classical three-stage method. Although the two- and one-stage methods lower the pH, the aroma and flavor of the doughs they produce are less intense than those of doughs made with the three-stage method. Low-viscosity pumpable doughs suitable for processing times of 1 week are made with special cultures by what is known as the "storable sourdough method" (Spicher and Stephan 1982).

The sourdough used for daily production in craft bakeries is made in kneading bowls. By contrast, the sourdough required for continuous bread production in industrial bakeries is frequently made in continuously operating units, usually in two stages (Fig. 7.2). Such units essentially comprise a thermostatically controlled

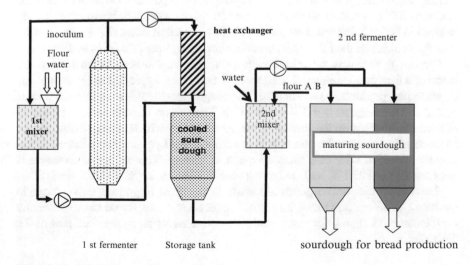

Fig. 7.2 Facility for the continuous production of sourdough

fermentation tube or cylindrical tank and a fermentation tank. The first stage involves the continuous propagation of the lactic acid bacteria, acid formation, and maturing. The fermented dough obtained in the first stage is divided, one part serving as an inoculum for the new sourdough, while the remainder is used in the second stage or cooled in a storage tank until further use. Acidification of the dough also promotes the swelling of the swelling materials (pentosans), and thus the water absorption of the dough, as well as improving the elasticity of the crumb by comparison with unleavened rye bread (Klingler 2010).

The flour used for making bread generally has a particle size lower than 180 μm. Meal is also used for making bread in Germany. It consists of partially ground cereal grain and may also contain the germ. The particle sizes generally range from 250 to 1400 μm. The particles, which are larger than flour particles, absorb water slowly and frequently insufficiently during dough production and may result in the bread being dry and crumbly. Prior to making the bread dough, part of the meal is therefore made into a sponge so that it has enough time to absorb water. There are three ways of doing this. Soakers are made by covering one part meal with one part water at around 20 °C and used to make bread dough after 16–24 h have elapsed. Scalders are made with one part water with a temperature of 60 °C mixed with two parts meal; swelling is complete after around 3–6 h. "Boiled sponges" are made by mixing one part meal with two parts water with a temperature of around 100 °C. In this case, the meal can be used after around 2 h. Meal doughs are soft doughs requiring little energy for kneading. Dough preparation as described for wheat doughs is neither necessary nor possible in this case. Doughs such as these can be baked in pans or molds without further processing (Freund 2009).

Pumpernickel is a German specialty which was first made in Westphalia several centuries ago. Pumpernickel was originally made solely of rye and rye meal. A scald is made with the cereal and allowed to swell for 8–10 h. In addition to sourdough, yeast, and salt, sugar beet syrup is added for color. The dough is partially baked in closed pans at around 200 °C and then baked for a further 16–24 h at a temperature that is gradually reduced to 100 °C. The bread is characterized by its lack of crust, slightly sweet taste, very dark color, and the fact that it stays fresh for a long time (Täufel et al. 1993a, b).

Lye dough products are traditionally produced in Southern Germany and are made of a firm dough prepared from white wheat flour. Typical shapes are pretzels or elongated products. After shaping, the dough pieces are left to rise for a brief period and then dipped in a 3 % lye solution. This causes the chemical degradation of flour constituents on the surface of the dough pieces which is responsible for the characteristic brown color and crispness of the crust. Lye dough products are frequently sprinkled with salt, caraway seeds, or cheese. The baking temperature is between 220 and 230 °C and the baking time is 12–15 min (Skobranek 1991).

Bread and other baked goods are some of the most important staple foods in Germany, thanks in part to the variety of products available. Bread has traditionally been eaten in Germany for many centuries and remains an established part of the daily diet.

References

Belitz H-D, Grosch W (1987) Lehrbuch der Lebensmittelchemie, 3. Springer, Berlin
Busch KG (2010) Verfahren Weizenbrotherstellung—Dosieren. In: Freund W (ed) Handbuch Backwaren Technologie. Behr's-Verlag, Hamburg, pp 1–29
Freund W (1995) Verfahrenstechnik Brot und Kleingebäck. Gildebuchverlag, Alfeld
Freund W (2009) Technologie der Vorteige. Behr's-Verlag, Hamburg
Goetz H (1973) Kinderlieder Kinderreime. Üeberreuter, Wien
Jacob HE (1954) Sechstausend Jahre Brot. Rowohlt Verlag, Hamburg
Klingler RW (2010) Grundlagen der Getreidetechnologie. Behr's-Verlag, Hamburg
Skobranek H (1991) Bäckereitechnologie. Verlag Handwerk und Technik, Hamburg
Spicher G, Stephan H (1982) Handbuch Sauerteig. BBV-Verlag, Hamburg
Täufel A, Ternes W, Tunger L, Zobel M (1993a) Lebensmittel-Lexikon A-K. Behr's-Verlag, Hamburg
Täufel A, Ternes W, Tunger L, Zobel M (1993b) Lebensmittel-Lexikon L-Z. Behr's-Verlag, Hamburg
Ternes W (1990) Naturwissenschaftliche Grundlagen der Lebensmittelzubereitung. Behr's-Verlag, Hamburg
Tscheuschner H-D (1996) Grundzüge der Lebensmitteltechnik, 2. Auflage Behr's-Verlag, Hamburg
Verband Deutscher Mühlen: Daten und Fakten 2009

Chapter 8
Production of Pastas with Bread Wheat Flour

C.S. Martínez, M.C. Bustos, and A.E. León

Contents

C.S. Martínez • M.C. Bustos • A.E. León (✉)
Instituto de Ciencia y Tecnología de los Alimentos Córdoba (CONICET—Universidad
Nacional de Córdoba), Córdoba, Argentina
e-mail: aeleon@agro.unc.edu.ar

© Springer Science+Business Media New York 2016
A. McElhatton, M.M. El Idrissi (eds.), *Modernization of Traditional Food
Processes and Products*, Integrating Food Science and Engineering Knowledge
Into the Food Chain 11, DOI 10.1007/978-1-4899-7671-0_8

8.1 Introduction

Pasta, a traditional food with high consumer acceptance because of its convenience, palatability, and nutritional quality, is consumed around the world (Petitot et al. 2009). There are those who prefer to make their own pastas with durum wheat flour using traditional methods, however, most tend to use commercially available pastas. Many factors contribute to pasta's popularity, especially its nutritional profile. It is a good source of complex carbohydrates and a moderate source of proteins and vitamins. For example, a 55 g portion of dry pasta contains about 210 cal from about 75 % carbohydrates. In the USA, dietary guidelines published by the United States Department of Agriculture and by Health Canada show that grain-based products, which include pasta, should be a major part of a healthy diet. Pasta has good consumer value and because of that it sells well in both good and bad economic times. Besides, dry packaged pasta is virtually non-perishable if stored appropriately, pasta is easy to cook, it has a wholesome taste, and an extensive variety of dishes can be prepared using the very different pasta shapes and sizes available nowadays (Marchylo and Dexter 2001).

8.2 Pastas

Approximately 12 million tons of pasta are produced worldwide per year. Italy is the world's largest pasta producer, with a production of 3,160,000 t a year (26 %). The USA with 2,000,000 t represents 16 % of worldwide production, and Brazil with 1,500,000 t (12 %) follows Italy. Considering pasta annual consumption, Italy is also the major consumer with 26 kg per person (Lezcano 2009).

The preferred ingredient to make traditional pasta is durum wheat semolina, obtained by milling durum wheat seeds (*Triticum durum* Desf.) Unfortunately, durum wheat grows under a relatively narrow range of climatic conditions; it cannot be cultivated in areas where the weather is too cold, too warm, or too wet.

In areas where durum wheat is expensive, supply and availability is low, and when pasta is produced in areas far from durum wheat cultivation areas, bread wheat (*Triticum aestivum* L.) flour is often blended with durum or used to produce pasta; the result is a good quality product, but not yellow in color and not as resistant to overcooking as pasta produced from durum wheat semolina (Hoseney 1994; Heneen and Brismar 2003).

Considering the limitations just mentioned, it is very common to find pasta made from bread wheat flour, especially in products consumed by people who prioritize price over quality. Even pasta made with bread wheat can be used to incorporate ingredients that improve nutritional quality (Bustos et al. 2011, 2013; Martínez et al. 2014; Rodríguez De Marco et al. 2014).

Pasta made with bread wheat flour has been compared with pasta made with durum wheat semolina (Martinez et al. 2007). To evaluate the quality of commercial spaghetti made with both types of wheat, cooking properties, optimum cooking

Table 8.1   OCT, cooking loss, water absorption, and amylose content in cooking water

Sample[a]	OCT (min)	Cooking loss (% w/w[b])	Water absorption (% w/w[b])	Amylose (% w/w[b])
S-Sem	10.5	5.3±0.0 b	266±1 c	2.0±0.2 a
S-TP1	8.5	4.4±0.0 a	256±1 ab	2.2±0.2 a
S-TP2	9.5	4.5±0.0 ab	254±3 a	2.5±0.2 b
S-TP3	7.5	6.4±0.7 c	259±0 b	4.7±0.1 d
S-TP4	7.5	6.3±0.3 c	267±0 c	3.7±0.1 c

Values followed by a different letter are significantly different ($P<0.05$)
Results were expressed as the mean of replications±SD. OCT, optimal cooking time (Martinez et al. 2007)
[a]S-Sem: sample made from durum wheat semolina, S-TP1 to S-TP4: commercial samples made from bread wheat flour
[b]Values expressed as g/100 g of dry pasta

time (OCT), cooking loss, water absorption, and amylose content in cooking water (Table 8.1) have been analyzed. Although in general samples made from bread wheat flour showed decreased OCT than samples made from durum wheat semolina (S-Sem), in some of them that difference was only slight.

Dick and Youngs (1988) established that cooking losses should be around 7 % and should not exceed 8 % for spaghettis made from durum wheat semolina, and water absorption values should be three times over the dry weight if a good-quality final product is expected. All pasta made from bread wheat flour and durum wheat semolina evaluated by Martinez et al. (2007) showed cooking losses lower than 7 g/100 g of dry pasta, while water absorption values were between 254 and 266 g/100 g of dry pasta; some samples made from bread wheat flour presented even lower values than S-Sem. In addition, amylose content in cooking water of bread wheat pasta showed similar values to S-Sem; this parameter gives information about the proportional amylopectin enrichment that takes place in the pasta surface, with higher values leading to an increase in adhesiveness.

Instrumental firmness was evaluated in cooked pasta at OCT and at 150 and 200 % of OCT (Fig. 8.1). Pasta made from durum wheat semolina and some made from bread wheat flour presented similar firmness at OCT, and a good resistance to over-cooking at 150 and 200 % of OCT (Martinez et al. 2007).

A sensory evaluation of pasta made from both types of wheat was also performed by Martinez et al. (2007). Yellow color, shininess, firmness, chewiness, elasticity, surface smoothness, and superficial defects were evaluated by the authors. Classifications obtained for all positive parameters by each judge were represented in a radial graph, where a higher area implies better pasta quality. Figure 8.2 shows that S-Sem and some samples of pasta made from bread wheat flour presented the highest areas, so these samples resulted in a better quality according to panel evaluation.

Both cooking and textural properties as well as sensory evaluation of pasta made from both types of wheat showed that the quality of pasta made from bread wheat flour can be similar to the quality of pasta made from durum wheat semolina (Martinez et al. 2007).

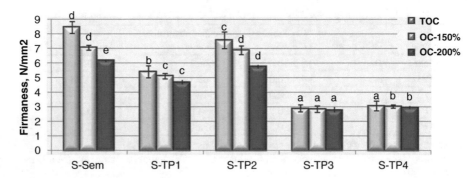

Fig. 8.1 Firmness of pastas cooked at optimum cooking time (OCT) and an overcooking of 150 and 200 % (OC-150 %, OC-200 %). Values followed by a different letter, at the same cooking time, are significantly different ($P < 0.05$) (Martinez et al. 2007)

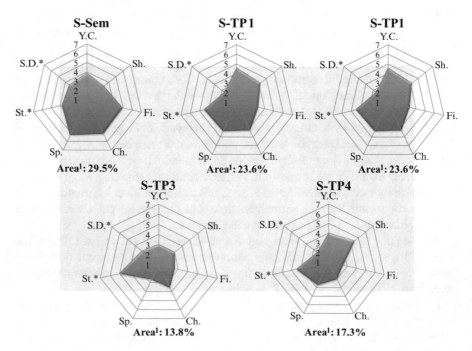

Fig. 8.2 Radial representation of commercial pasta sensory evaluation. S-Sem: sample made from durum wheat semolina, S-TP1 to S-TP4: samples made from bread wheat flour. Y.C. Yellow color, Sh.: shininess, Fi.: firmness, Ch.: chewiness, Sp.: springiness, S.D.: superficial defects, St.: stickiness. *: opposite values were used for surface smoothness and superficial defects in order to include them in the graphic with a positive attribute. 1: Area percentage from total graphic area (Martinez et al. 2007)

8.3   Raw Material

Scientific research has been undertaken to understand the parameters influencing industrial pasta processing and the quality of the final product. The choice of raw material and processing variables in the production of dry pasta are the only measures one can take to ensure *al dente* characteristics, namely, a firm and resilient pasta with no surface stickiness and little if any cooking losses (Brunnel et al. 2010).

8.3.1   Wheat Flour

White wheat flour is obtained from grain endosperm. Considering the groove present in the grain, it is impossible to remove the outer layers by simple abrasion. For this reason, successive grindings (called molturations) followed by sieving are carried out to separate the different grain fractions corresponding to teguments and aleurone layers (bran), germ, and endosperm (Cheftel and Cheftel 1992).

Wheat flour has an important role in pasta processing as this ingredient accounts for 95–98 % of solids in pasta.

Starch is the major component of wheat flour (approximately 65 g/100 g flour). Starch consists of two main structural components, amylose, which is essentially a linear polymer, and amylopectin, which is a larger branched molecule. Starch granules are affected by thermal treatment due to their partially crystalline native structure, so the starch granule experiments phase transitions called gelatinization and retrogradation (Belitz and Grosch 1999).

Wheat flour proteins represent 10–15 % of total flour and can be divided into two groups: gluten proteins and non-gluten proteins. The first ones are storage proteins and correspond to 75–80 % of total proteins, while non-gluten proteins (20–25 % of total proteins) include the majority of enzymes.

Lipids (approximately 2 % of flour), minerals (0.5 % of flour expressed as ash weight), enzymes (α-y β-amylases), non-starch polysaccharides (e.g., pentosans), and pigments like carotenes and flavonoids, which are responsible for flour color that correlates with pasta color, are present in small quantities in wheat flour. Finally, water content of wheat flour is approximately 14 g/100 g of flour (Ribotta et al. 2009).

A relatively fine flour particle size enables even hydration during mixing and an optimum and uniform gluten development during sheeting. Typical noodle flour retains less than 15 % of material on 100 μm. The particle size distribution should be uniform because small flour particles hydrate much faster than big ones, generating dough particles of different sizes that form spots (wet or dry) on the dough sheet. Flour with very fine particle size may be indicative of high starch damage, which should be avoided, due to its competition for water with gluten during mixing (Fu 2008).

8.3.1.1 Importance of Protein in Pasta

The unique ability of wheat flour to form cohesive, elastic and extensible dough is due to the presence of gluten proteins. With regard to this, the rheological properties of dough are controlled by proteins (De Noni and Pagani 2010), and the quantity and quality of these proteins are important in pasta processing (Hoseney 1994). Pasta with an elastic and chewy texture is produced by high concentrations of proteins (10–14 %) which are able to form a strong gluten matrix. On the other hand, pasta made from flour with low concentrations of proteins has a low resistance to cooking and become soft and sticky. In consequence, adequate protein content is important to textural properties (Ross et al. 1997; Park and Baik 2004; Zhao and Seib 2005).

It is remarkable that dried pasta generally contains higher protein content than fresh pasta, because the dried pasta has to be able to withstand the drying process without breaking (Fu 2008). In relation to this, pasta quality is related not only to protein content, but it also correlates with gluten strength (D'Egidio et al. 1990; Malcomson et al. 1993; Rao et al. 2001).

In the quality evaluation of wheat flour proteins, individual protein fractions should be considered (Edwards et al. 2003; Sissons et al. 2005). Glutenins and glia-dins are responsible for good cooking properties, since the protein matrix tenacity and elasticity are determined by protein/protein and subunit/subunit aggregates. Besides, other chemical properties of proteins, like sulfhydryl groups or low molecular weight glutenin content, are related to pasta quality. In order to evaluate wheat quality for pasta processing, other additional parameters like glutenin/gliadin, the presence of specific protein fractions, superficial hydrophobicity, and functional properties of gluten and dough should be considered (De Noni and Pagani 2010).

8.3.1.2 The Importance of Starch in Pasta

The role of starch in the rheological properties of dough for making pasta has been underestimated if compared to the attention gluten has received, because the super-ficial characteristics of granules can affect dough viscoelastic behavior, since this characteristic determines the type of protein–starch interactions (De Noni and Pagani 2010). According to Fu (2008), pasta made with high swelling starch flour has softer texture than those with low swelling starch.

In pasta processing, during mixing the temperature is under 50 °C and starch granules absorb water slowly due to the presence of a proteins-phospholipids layer that limits swelling and gelatinization. In consequence, starch granules that have a high temperature of gelatinization, delaying swelling and solubilization, are good for pasta processing because these properties reduce the interferences with protein matrix development. The presence of gluten increases the starch gelatinization temperature (De Noni and Pagani 2010). A high proportion of small starch granules (5–10 μm) (Soh et al. 2006) and a high amylose/amylopectin ratio have been proposed to have the same effect. Amylopectin is the starch component related to high stickiness in cooked pasta (De Noni and Pagani 2010).

Besides, the mechanical breaking of starch granules (damaged starch) that takes place during milling and the action of enzymes like α-amylase should be considered negative modifications in starch (Matsuo et al. 1982), as they promote starch solubilization during pasta cooking (De Noni and Pagani 2010). An increase in damaged starch affects pasta color negatively, promoting an increase in cooking losses and excessive swelling of pasta surface (Hatcher et al. 2002).

8.3.2   Water

Water is the second most important raw material after flour in pasta manufacturing. Water provides the necessary medium for all the physicochemical and biochemical reactions that underlie the transformation of raw materials into a finished product. Without water, gluten proteins in the flour cannot exhibit viscoelastic properties. Water-soluble ingredients are usually dissolved in water before mixing. However, the amount of water required for pasta processing needs to be optimized so that it can hydrate the flour and allow the development of a uniform dough sheet, and yet prevent handling or sheeting problems in the formed dough due to stickiness. Water absorption level for pasta processing is about 30–38 % based on flour weight (Fu 2008).

Also, the quantity of water added to flour for mixing affects pasta color. In a previous study, it was found that the lightness of fresh pasta made from different varieties of bread wheat flour (Baguette, BAG, Buk Guapo, BUK, and commercial wheat flour) and triticale flour (*Triticosecale Wittmack* var. Tatú) was negatively affected by increased water addition during mixing (Fig. 8.3a). In relation to this, b* and a* increased with increasing water addition (Fig. 8.4b, c) (Martinez et al. 2012). Similar results had been observed in previous research in pasta (Hatcher et al. 1999; Humphries et al. 2004; Wang et al. 2004; Solah et al. 2007; Ohm et al. 2008).

Apart from the basic fundamental sanitary requirements, water used for pasta processing has to meet certain specifications in order to produce high quality

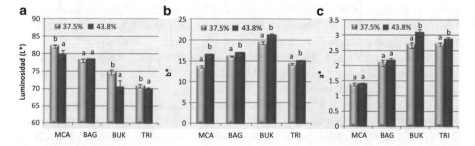

Fig. 8.3 Effect of addition of different quantities of water during mixing in L* (**a**), b* (**b**), and a* (**c**) parameters of fresh pasta made with 37.5 and 43.8 mL of water per 100 g of flour. Bars with different letters for the same sample are significantly different ($P<0.05$), (Martinez et al. 2012)

Fig. 8.4 Drying defect:
strands division in sheeted
pasta made from bread
wheat flour. Dry pasta (**a**)
and cooked pasta (**b**) (data
not published)

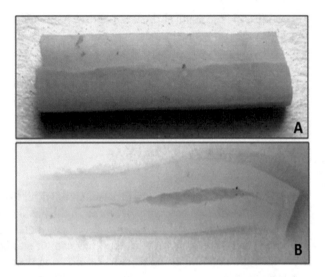

products. Water varies in hardness, alkalinity, and pH value, which in turn affects
flour hydration, dough sheet properties, starch gelatinization, and texture of the
finished products. Excessively hard water is undesirable because it retards flour
particles hydration by tightening the gluten proteins. The ions present in water also
have a very significant impact on the gelatinization of starch during steaming or
boiling. On the other hand, very soft water is objectionable since it lacks the gluten-
strengthening minerals and tends to yield soft, sticky dough sheets. Water of medium
to low hardness is considered suitable for noodle processing (Fu 2008).

8.3.3 Salt

The amount of salt added is usually 1–3 % of flour weight. Salt performs three principal
functions in noodle processing. The most important is the strengthening and tightening
effect on gluten and the improvement of viscoelastic properties (Dexter et al. 1979),
which is partly due to its inhibitory effect on proteolytic enzymes, although other
evidence indicates a more direct interaction of salt with flour proteins (Fu 2008).

A second function of salt is flavor enhancing and texture improving effects. Pasta
with added salt has a shorter cooking time (Dexter et al. 1979) and a softer but more
elastic texture than that without salt.

A third function of salt is the inhibition of enzyme activities and the growth of
microorganisms. Salt slows down the alcoholic and lactic fermentation process.
When making dried pasta, the amount of salt in the noodle can affect the rate of
drying. Moisture evaporates more slowly in pasta with higher amounts of salt,
preventing fractures during drying cycles (Dexter et al. 1979).

8.4   Pasta Processing

According to the type of process, pasta is classified as sheeted (ribbon-cut pasta) or extruded (spaghettis). Although pasta can have many variations in formulation, shape, and size, the process of making pasta is quite constant.

8.4.1   Mixing

In pasta manufacturing, the main aims of mixing are to distribute the ingredients uniformly and to allow hydration of flour particles. There is little gluten development during the mixing stage in the low water absorption pasta dough in addition to very short mixing times. The degree of gluten development, however, can be very significant in high water absorption dough, as in bread processing (~50 %) with a long mixing time (De Noni and Pagani 2010).

8.4.2   Dough Resting

Mixing is usually followed by dough resting. This step allows crumbly mixture to accelerate further hydration of flour particles and to redistribute water in the dough system. Resting can also improve processing properties and facilitate gluten formation during sheeting, which is achieved by the relaxation of the gluten structure already formed during mixing (Fu 2008).

8.4.3   Sheeting

Although flour particles are sufficiently hydrated after mixing and resting, the development of the gluten matrix is far from complete and is localized without continuity. It is during the sheeting process that the continuous gluten matrix is developed. Under compression, adjacent endosperm particles become fused together so that the protein matrix within one flour particle becomes continuous with the adjacent particles. The sheeting process is intended to achieve a smooth dough sheet with the desired thickness, and a continuous and uniform gluten matrix in the dough sheet.

In general, it is accepted that sheeted pasta should have a better quality compared with extruded pasta, as gluten network reaches a greater development during sheeting than during extrusion (Matsuo et al. 1978; Dexter et al. 1979).

8.4.4 Cutting

Once the dough sheet is reduced to the desired thickness, the sheet is then cut into noodle strands along the direction of sheeting. The width and shape of the noodle strands are determined by cutting rolls.

In the case of extrusion, when pasta comes out of the extruder screw and reaches the head, which is generally rectangular for long pasta and circular for short pasta, dough is cut with a blade. Inserts, either made of Teflon or bronze, are located inside circular heads and determine the various pasta shapes according to their design (Calvelo 2008).

8.4.5 Drying

The shelf life of pasta can be significantly extended if the microbiological and bio-chemical stability is ensured. The most effective way of achieving this goal is to dry pasta to a moisture content at which microbiological growth is impossible. An adequate drying process involves many stages in order to minimize undesirable structural changes. A very usual practice is a drying process with three stages: pre-drying, drying, and cooling. The first stage, which takes up to 15 % of total drying time, is of primary importance. In this stage, pasta moisture content is reduced from 32 to 38 % to less than 28 %. Its main function is to dry the noodle superficially soon after cutting to prevent noodle strands from sticking together and to avoid the over-elongation of pasta strands (Calvelo 2008). The preservation of pasta capillarity is essential for water redistribution at the following stage (Professional Pasta, L1N06P044).

The following drying phase must include alternating phases of water evaporation from the surface and inner redistribution. The speed of this phase is inevitably slower than that of pre-drying because the structure of the product has become more rigid, capillary action has decreased and so the migration of the remaining particles of water from the inside to the outside of the product is slower. Drying normally takes approximately 6–8 times longer than the time required for pre-drying (Professional Pasta, L1N06P044).

To avoid excessive tension inside the product structure, all the drying process must be interspersed with tempering phases, i.e., periods of minimum air circulation and high humidity, to allow water diffusion from the center to the pasta surface (Calvelo 2008).

An inappropriate drying process can damage pasta structure and generate over-elongation, cracks, deformation, and strands division (Fig. 8.4) which causes many problems during pasta manipulation and packing.

There are three different technologies for drying: drying at low temperature (LT) (<60 °C), drying at high temperature (HT) (60 °C<T<90 °C), and drying at very high temperature (VHT) (T>90 °C). The application of high temperature can be performed under two conditions: high temperature-high humidity (HT-HM) (70–75

°C, 20–25 % H) or high temperature-low humidity (HT-LM) (70–75 °C, ~18 % H). In general, high temperatures reduce drying times and increase the process capacity. Besides this, microbiological quality and cooking properties are improved and yellow color is favored. However, high temperatures can have detrimental effects on the nutritional value of pasta due to the decrease in available lysine (Maillard reaction), and can also increase an undesirable red color in pasta (Calvelo 2008).

8.4.6   Cooking Properties of Pasta

The cooking properties of dry pasta are the result of the characteristics of raw materials and processing conditions.

Pasta is a system with limited humidity; during cooking, a strong competition for water between starch and proteins takes place. Proteins need water for coagulation and produce an elastic matrix; at the same time, starch swells, gelatinizes, and becomes more soluble due to water absorption. Because proteins denaturation and starch swelling occur approximately at the same temperature, there is a physical competition between these two processes during cooking. When interactions among proteins prevail, starch, which hydrates slowly, remains trapped in the protein matrix and cooked pasta will be firm and surface adhesiveness will be low, preventing the strands to stick together. On the other hand, when the protein matrix is not sufficiently strong and elastic, starch swells and gelatinizes before the coagulation of proteins occurs. In this case, amylose diffuses into cooking water and amylopectin remains on the surface, resulting in a soft and sticky texture (De Noni and Pagani 2010).

The key factors for cooking pasta are the relationship between water and pasta, the OCT, and the quality of cooking water. The desirable volume of boiling water is 10–20 times the weight of uncooked wet noodles (Fu 2008).

Pasta quality is expressed in terms of water absorption, leached material during cooking, and such texture properties as firmness and stickiness. The texture of cooked pasta is generally recognized as its most important quality aspect (Brunnel et al. 2010).

In a recent study, fresh pasta made from bread wheat flour was substituted with 5 and 10 % w/w of starch and 3 and 6 % w/w of gluten in order to evaluate how the main flour components influence pasta quality (data not published). Pasta with increased gluten content, G3 and G6, presented better cooking properties, according to increased OCT, decreased cooking losses and water absorption observed in these samples (Table 8.2). The increase in gluten content favors the formation of a firm structure, limiting water diffusion to the center of pasta and in consequence decreasing lixiviation of solid to cooking water. Other researchers (Zweifel et al. 2003; De Noni and Pagani 2010) found similar results.

The structure of two different types of pasta, one made from wheat flour substituted with 10 % w/w of starch (A) and the other substituted with 6 % w/w of gluten (B) after 3 min of cooking, is shown in Fig. 8.5. In the sample substituted with starch, there is a higher degree of starch gelatinization (translucent zone) than in

Table 8.2 TOC, Cooking loss, and water absorption in pastas

Sample[a]	TOC (min)	Cooking loss (% w/w)	Water absorption (% w/w)
Control	12	6.5±0.2 b	149±4 b
A5	12	6.7±0.0 bc	150±1 b
A10	12	6.8±0.1 c	152±1 b
G3	14	6.0±0.0 a	145±0 ab
G6	14	5.9±0.1 a	139±6 a

Values followed by a different letter are significantly different ($P<0.05$)
[a]Control sample: only made with wheat flour without any incorporation. A5 y A10: pasta made from flour substituted with 5 and 10 % of starch, G3 y G6: pasta made from flour substituted with 3 and 6 % of gluten (data not published)

Fig. 8.5 Structure of sheeted pasta made from wheat flour substituted with 10 % w/w starch (**a**) and 6 % w/w gluten (**b**) after 3 min of cooking (data not published)

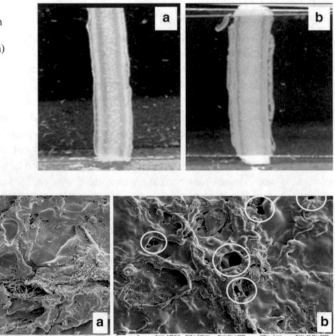

Fig. 8.6 Scanning electron microscopy of cooked pasta surface made from bread wheat flour substituted with 6 % w/w of gluten (**a**) and 10 % w/w of starch (**b**) (data not published)

the sample substituted with gluten, which indicates that the structure of the former was weaker and allowed an increase in the diffusion of cooking water into the center of pasta.

A more compact structure is observed in pasta with added gluten than in pasta with added starch, in scanning electron microscopy (SEM) photographs (Fig. 8.6a). Starch-enriched pasta (Fig. 8.6b) presents many pores on the cooked pasta surface which facilitate water diffusion to pasta, favors starch gelatinization, and thus its lixiviation to cooking water (data not published).

Table 8.3 Sensory evaluation of pastas made from bread wheat flour substituted with starch and gluten

Sample[a]	Firmness	Chewiness	Adhesiveness	Yellow color
Control	0 b	0 bc	0 bc	0 a
A5	−1 a	−1 ab	0 ab	−1 a
A10	−1 a	−1 a	−1 a	−1 a
G3	1 c	1 cd	1 c	1 b
G6	2 d	2 d	0 bc	2 c

Values followed by a different letter are significantly different ($P<0.05$)
[a]Control sample only made with bread wheat flour without any incorporation. A: Starch, G: gluten (data not published)

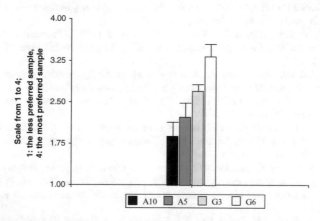

Fig. 8.7 Preference test of fresh pasta made from bread wheat flour substituted with 5 and 10 % of starch and 3 and 6 % of gluten (data not published)

Sensory evaluation using a multiple discriminative test for fresh commercial pasta made from bread wheat flour substituted with starch and gluten separately showed that the addition of 3 and 6 % w/w of gluten increases firmness and chewiness and decreases adhesiveness, while starch addition generates the opposite effect compared to control sample (Table 8.3). In this regard, starch-enriched pasta presented a decrease in yellow color, while gluten-enriched pasta showed an increase of this parameter, according to b* values determined by spectrophotometry of reflectance (data not published).

In this sense, when a preference test using a scale from 1 to 4, being 1 the less preferred sample and 4 the most preferred one, was carried out, sample G6 was the one with the best qualification while the sample with high percentage of starch substitution, A10, was the less preferred one (Fig. 8.7). Certainly, the consumer preference in pasta texture varies around the world, the first thing that the consumer takes into account when pasta quality is evaluated is its "*al dente*" texture because of Italian tradition, which is characterized by high firmness and low adhesiveness (data not published).

References

Belitz H, Grosch W (1999) Polysaccharides. In: Belitz H, Grosch W (eds) Food chemistry, 2nd edn. Springer, Berlin, pp 237–318

Bustos MC, Pérez G, León AE (2013) Combination of resistant starches types II and IV with minimal amount of oat bran yields good quality, low glycaemic index pasta. Int J Food Sci Technol 48:309–315

Bustos MC, Pérez G, León AE (2011) Sensory and nutritional attributes of fibre-enriched pasta. LWT-Food Sci Technol 44:1429–1434

Brunnel C, Pareyt B, Brijs K, Delcour JA (2010) The impact of the protein network on the pasting and cooking properties of dry pasta products. Food Chem 120:371–378

Calvelo A (2008) Material del curso Tecnología de elaboración de pastas secas. Universidad Nacional de La Plata. Maestría en tecnología de los alimentos

Cheftel JC, Cheftel H (1992) Granos vegetales. In: Acribia Z (Ed.), Introducción a la bioquímica y tecnología de los alimentos. España, pp 101–104

D'Egidio MG, Mariani BM, Nardi S, Novaro P, Cubadda R (1990) Chemical and technological variables and their relationships: a predictive equation for pasta cooking quality. Cereal Chem 67:275–281

De Noni I, Pagani MA (2010) Cooking properties and heat damage of dried pasta as influenced by raw material characteristics and processing conditions. Crit Rev Food Sci Nutr 50:465–472

Dexter JE, Matsuo RR, Dronzek BL (1979) A scanning electron microscopy study of Japanese noodles I. Cereal Chem 56:202–208

Dick JW, Youngs VL (1988) Evaluation of durum wheat, semolina and pasta in the United States. In: Fabriani G, Lintas C (eds) Durum wheat chemistry and technology. American Association of Cereal Chemists, St. Paul, pp 237–248

Edwards NM, Mulvaney SJ, Scanlon MG, Dexter JE (2003) Role of gluten and its components in determining durum semolina dough viscoelastic properties. Cereal Chem 80:755–763

Fu BX (2008) Asian noodles: history, classification, raw materials, and processing. Food Res Int 41:888–902

Hatcher DW, Anderson MJ, Desjardins RG, Edwards NM, Dexter JE (2002) Effects of flour particle size and starch damage on processing and quality of white salted noodles. Cereal Chem 79:64–71

Hatcher DW, Symons SJ, Kruger JE (1999) Measurement of the time dependent appearance of discolored spots in alkaline noodles by image analysis. Cereal Chem 76:189–194

Heneen WK, Brismar K (2003) Structure of cooked spaghetti of durum and bread wheats. Starch 55:546–557

Hoseney RC (1994) Pasta and noodles. In: Hoseney C (ed) Principles of cereal science and technology, 2nd edn. American Association of Cereal Chemists, St. Paul, pp 321–334

Humphries JM, Graham RD, Mares DJ (2004) Application of reflectance colour measurement to the estimation of carotene and lutein content in wheat and triticale. J Cereal Sci 40:151–159

Lezcano E (2009) Cadenas alimentarias: pastas alimenticias. Alimentos Argentinos 46:20–28

Malcomson LJ, Matsuo RR, Balshaw R (1993) Textural optimization of spaghetti using response surface methodology: effects of drying temperature and durum protein level. Cereal Chem 70:417–423

Marchylo BA, Dexter JE (2001) Pasta production. In: Owens G (ed) Cereals processing technology, 1st edn. Woodhead Publishing, Inglaterra, pp 1–3 (Chapter 6)

Martinez CS, Ribotta PD, León AE, Añón MC (2007) Physical, sensory and chemical evaluation of cooked spaghetti. J Texture Stud 38:666–683

Martinez CS, Ribotta PD, León AE, Añón MC (2012) Colour assessment on bread wheat and triticale fresh pasta. Int J Food Prop 15(5):1054–1068

Martínez CS, Ribotta PD, Añón MC, León AE (2014) Effect of amaranth flour (Amaranthus mantegazzianus) on the technological and sensory quality of bread wheat pasta. Food Sci Technol Int 20:127–135

Matsuo RR, Dexter JE, Dronzek BL (1978) Scanning electron microscopy study of spaghetti processing. Cereal Chem 55:744–753

Matsuo RR, Dexter JE, MacGregor AW (1982) Effect of sprout damage on durum wheat and spaghetti quality. Cereal Chem 59:468–472

Ohm JB, Ross AS, Peterson CJ, Ong YL (2008) Relationships of high molecular weight glutenin subunit composition and molecular weight distribution of wheat flour protein with water absorption and color characteristics of noodle dough. Cereal Chem 85:123–131

Park CS, Baik BK (2004) Cooking time of white salted noodles and its relationship with protein and amylose contents of wheat. Cereal Chem 81:165–171

Petitot M, Abecassis J, Micard V (2009) Structuring of pasta components during processing: impact on starch and protein digestibility and allergenicity. Trends Food Sci Technol 20:521–532

Professional Pasta (2008) General comment on pasta investigations into drying pasta. L1N06P044. www.professionalpasta.it, consultado en marzo de 2008

Rao VK, Mulvanney SJ, Dexter JE, Edwards NM, Peressini DJ (2001) Stress-relaxation properties of mixograph semolina-water doughs from durum wheat cultivars of variable strength in relation to mixing characteristics, bread- and pasta-making performance. J Cereal Sci 34:215–232

Ribotta PD, Pessoa Filho P, Tadini CC (2009) Masas congeladas. In: Ribotta PD, Tadini CC (eds) Alternativas Tecnológicas para la Elaboración y la Conservación de Productos Panificados, 1st edn. Universidad Nacional de Córdoba, Facultad de Ciencias Exactas, Físicas y Naturales, Argentina, pp 18–19 (Chapter 1)

Rodríguez De Marco E, Steffolani ME, Martínez CS, León AE (2014) Effects of spirulina biomass on the technological and nutritional quality of bread wheat pasta. LWT—Food Sci Technol 58:102–108

Ross AS, Quail KJ, Crosbie GB (1997) Physicochemical properties of Australian flours influencing the texture of yellow alkaline noodles. Cereal Chem 78:814–820

Sissons MJ, Egan NE, Gianibelli MC (2005) New insights into the role of gluten on durum pasta quality using reconstitution method. Cereal Chem 82:601–608

Soh HN, Sissons MJ, Turner MA (2006) Effect of starch granule size distribution and elevated amylose content on durum dough rheology and spaghetti cooking quality. Cereal Chem 83:513–519

Solah VA, Crosbie GB, Huang S, Quail K, Sy N, Limley HA (2007) Measurement of color, gloss, and translucency of white salted noodles: effects of water addition and vacuum mixing. Cereal Chem 84:145–151

Wang C, Kovacs MIP, Fowler DB, Holley R (2004) Effects of protein content and composition on white noodle making quality: color. Cereal Chem 81:777–784

Zhao LF, Seib PA (2005) Alkaline-carbonate noodles from hard winter wheat flours varying in protein, swelling power, and polyphenol oxidase activity. Cereal Chem 82:504–516

Zweifel C, Handschin S, Escher F, Conde-Petit B (2003) Influence of high-temperature drying on structural and textural properties of durum wheat pasta. Cereal Chem 80:159–167

Part II
Americas and the Rest of the World

Chapter 9
Use of Edible Coatings, a Novel Preservation Method for Nuts

Lorena Atarés, Amparo Chiralt, and Anna McElhatton

Contents

9.1 Introduction

It is widely known that nuts have high lipid content and are rich in essential fatty acids (especially linoleic and linolenic acids). Their fatty profile very rich in mono- and polyunsaturated fats makes them very prone to lipid oxidation and rancidity. Nonenzymatic oxidation in roasted peanuts is known to be the major cause of rancidity, since high temperature eliminates the activity of lipoxygenase (Lee et al. 2002). In

L. Atarés (✉) • A. Chiralt
Instituto de Ingeniería de Alimentos para el Desarrollo,
Universidad Politécnica de Valencia, Valencia, Spain
e-mail: loathue@tal.upv.es

A. McElhatton
Faculty of Health Sciences, University of Malta, Msida, Malta

© Springer Science+Business Media New York 2016
A. McElhatton, M.M. El Idrissi (eds.), *Modernization of Traditional Food Processes and Products*, Integrating Food Science and Engineering Knowledge Into the Food Chain 11, DOI 10.1007/978-1-4899-7671-0_9

their natural state, nuts have natural built-in packaging protection in the form of skins and shells (Miller and Krochta 1997). These natural barriers retain flavor and aroma and regulate the oxygen transport, as well as that of carbon dioxide and moisture.

However, processed foods that lack these natural barriers unfortunately dominate modern dietary choices. Therefore, in minimally processed nuts, the kinetics of lipid oxidation has to be controlled if the development of off-flavors that make the product unacceptable to the consumer is to be controlled. This is of significance because nuts in general and almonds especially are main ingredients of many traditional confectionery products such as *turrón* (a form of nougat) or marzipan, and any degradation of nut quality would consequently lead to loss of product quality.

Nuts are also ingredients in many dishes and widely consumed as snacks. Indeed, nuts are considered a healthy food choice when they form part of a balanced and healthy diet because they have proven cardioprotective effects. According to Abbey et al. (1994), replacing half of the daily fat intake with nuts has been known to lower total and LDL cholesterol levels in humans significantly. Numerous clinical studies have revealed that low-density lipoprotein cholesterol reductions of 10–15 % have been observed, where walnuts, almonds, macadamias, hazelnuts, pecans, or peanuts were incorporated into the diet (Kurlandsky and Stote 2006). Over decades, synthetic plastics (petrochemical based) have been the traditional materials used for the packaging of food products. A variety of synthetic polymers and laminates have been developed and represent an excellent barrier to oxygen transference (Miller and Krochta 1997). The increased use of synthetic packaging films has resulted in serious ecological problems due to their non-biodegradability (Tharanathan 2003). The concern for a safe environment has led to a shift towards the use of biodegradable materials, i.e., edible films and coatings.

9.2   Edible Films and Coatings to Maintain Quality and Safety

In the last few years, the research in the field of edible films and coatings has attracted significant attention within the scientific community. The main advantage of these alternative materials is the reduction of synthetic packaging and the increase of recyclability, while having the potential to limit moisture, aroma, and lipid migration between food components (Miller and Krochta 1997; Nussinovitch 2009).

An edible film or coating consists of a thin layer of an edible material that is applied on the food product and is able to protect it from the environment. When formulating an edible film, at least one of the components must be capable of forming a structural matrix with a sufficient cohesiveness (Debeaufort et al. 1998). This layer should provide a good moisture barrier so various materials such as carbohydrates, proteins, lipids, or hydrocolloids have been suggested and were tested, and found to vary in their suitability. Carbohydrates and proteins were tried but found insufficient due to their hydrophilic nature; however, this issue was mitigated through the addition of lipids to the film formulation, to make composite films (Tharanathan 2003). These multicomponent edible films and coatings have the advantage of possessing the complementary desirable functional properties of each their components and counterbalance any component shortcomings. A composite film formulation can be tailor-made to suit the needs of a specific commodity (Tharanathan 2003). Composite films are generally based on a hydrocolloid structural matrix to which a plasticizer (such as sorbitol or glycerol) is added to promote the formation of edible films and coatings with good mechanical properties. Many materials are too brittle and would not form proper films without the addition of these agents. Brittleness is reduced through the changes in the hydrogen bonding between the film polymers.

An additional advantage of edible films and coatings is that they could act as vehicles for edible active ingredients that could be added to the formulations. The film material could be used to encapsulate selected additives which may have product enhancing or functionality such as an antioxidant action. Other possibilities include the encapsulation of antimicrobial agents, aroma compounds, and pigments, ions that stop browning reactions or nutritional substances such as vitamins (Debeaufort et al. 1998).

The choice of a film-forming substance, and additives, depends mainly on the specific characteristics of the food product to be coated. In general, the most important general prerequisites to be fulfilled by edible films and coatings are the following (Debeaufort et al. 1998):

• Good sensory qualities
• High barrier and mechanical properties
• Enough biochemical, physicochemical, and microbial stability
• Free of toxics and safe for health
• Simple technology
• Nonpolluting
• Low cost of raw materials and process

In the last decade, the research into the use of edible films and coatings to maintain quality and extend the shelf life of food products has attracted significant attention. They have been successfully used in the shelf-life extension of many different products: meat and meat derivatives (Oussalah et al. 2004; Chidanandaiah and Sanyal 2009; Ou et al. 2005), fish and derivatives (Gómez-Estaca et al. 2007; Duan et al 2010; Lopez-Caballero et al. 2005), fruits (Pérez-Gago et al. 2006; Tapia et al. 2008), and vegetables (Ponce et al. 2008; Ayranci and Tunc 2003).

9.3   Application of Tailor-Made Edible Coating to Nuts

Preservation of nuts through the use of corn zein and shellac wax coatings has been used for hundreds of years (Tharanathan 2003). According to Debeaufort et al. (1998), in the nineteenth century, sucrose was initially applied as an edible coating on nuts, almonds, and hazelnuts to prevent oxidation and rancidity during storage. In recent years, a variety of materials have been tested in order to meet the specific requirements of this type of products. In the search for suitable film-forming materials for the proper coating of nuts, the following requisites should be taken into account.

9.3.1 Organoleptic Compatibility

Being considered as food components, edible films and coatings are usually required to be as tasteless as possible in order not to be detected during the consumption (Contreras-Medellin and Labuza 1981). When applied to nuts, edible coatings should not modify the taste and flavor of the product. If this is not possible, the organoleptic properties of the coating should be compatible with that of the nut (Biquet and Labuza 1988).

9.3.2 Antioxidant Activity

Given that nuts are rich in oxidation-sensitive lipids, the coating material should be able to protect these components from the oxidation process that would result in off-flavors. Edible films and coatings that are able to do so, are claimed to have antioxidant activity. This protective capability of the edible films and coatings is the result of two different actions:

(a) Barrier to oxygen. Among all the factors affecting the rate of lipid oxidation, oxygen concentration is one of the most important (Labuza 1971). If oxygen availability is reduced by the barrier action of the coating, the oxidation rate is diminished. For this reason, materials with low oxygen permeability are the best choice in order to obtain the best coatings for oxidation-sensitive products such as nuts (Nussinovitch 2009; Swenson et al. 1953). In this respect, ambient relative humidity plays an important role in the stability of the product because the oxygen permeability of edible films and coatings depends greatly on water availability. The greater the relative humidity, the greater the water content in the film, which promotes molecular mobility and diffusion controlled processes, such as oxygen permeation (Hong and Krochta 2006).

(b) Specific action of antioxidant additives. As commented on above, edible films and coatings are able to encapsulate active ingredients or additives exhibiting some specific action in the system. The incorporation of antioxidants is recommended in our context (Nussinovitch 2009; Cosler 1957), given that these chemicals will have an antioxidant action on the product. In the last few years, natural antioxidants have received a great deal of attention because of the worldwide trend to avoid synthetic food additives. According to Frankel (1996), natural antioxidants in food products may have clear benefits because they have anticarcinogenic effects and inhibit biologically harmful oxidation reactions in the body. Due to their antioxidant and antimicrobial properties, essential oils are being studied as additives to be incorporated into edible films and coatings (Atarés et al. 2010).

9.3.3 Good Adhesiveness on the Surface of the Nut

The antioxidant effect of the coating will be effective so long as the coating is in close contact with the nut surface. Evidently, the occurrence of cracks and flaking in the coating would drastically affect its barrier properties. Generally speaking, the adhesiveness of the coating material on the food product is highly dependent on the nature of both, as well as their affinity to each other. Adhesiveness depends on the nature and on the number of interactions between film and support (Debeaufort et al. 1998).

In the case of nuts, the inherently poor adherence of coating (hydrophilic) and nut surface (hydrophobic) may be the cause of incomplete coverage, which is a common and important setback. According to Lin and Krochta (2005), this problem could arise during the coating step (dewetting), the drying (shrinking and cracking), and the storage of the coated product (flaking). The addition of a surfactant (aka emulsifiers or surface active agents) to the coating formulation is a common technique to solve this problem. They allow adherence and mixing between hydrophobic and hydrophilic materials. Their incorporation would reduce the interface tension between coating and solid, hence the interactions between both are promoted and the extensibility is improved. An alternative solution would imply the modification of the nut surface with surfactant adsorption prior to the coating. Both techniques will be commented on in Sect. 9.4.

9.3.4 Microbial Stability

Contamination of nuts by aflatoxins is a serious health problem. These toxins are produced from infection by fungus *Aspergillus* and are considered as one of the most dangerous contaminants of foods (Basaran et al. 2008; Luttfullah and Hussain 2011). The incorporation of an antimicrobial agent should also be considered in the formulation of coatings for nuts.

9.3.5 Good Optical Properties

An additional interesting requisite for nut coating materials would be their ability to improve the general appearance of the coated product. In this respect, the color of the nuts should not be altered and the coatings should exhibit high transparency. The gloss of the nuts can be improved to make them more appealing. Water-soluble cellulose derivatives (methylcellulose (MC), hydroxypropyl cellulose (HPC), carboxymethyl cellulose (CMC), and hydroxypropyl methylcellulose (HPMC)) produce transparent and shiny films (Debeaufort et al. 1998).

Table 9.1 Original research studies—performed over the last two decades on the application of edible coatings to nuts

Nut product	Coating		Reference
Peanuts	Whey protein	Sugar esters, soy lecithin, Sorbitan laureate (Span 20) (*s*)	Lin and Krochta (2005)
		Glycerol (*p*)	Lee et al. (2002)
		Lecithin (*s*)	
		Methyl paraben (*am*)	
		Vitamin E (*ao*)	Lee and Krochta (2002)
		Glycerol (*p*)	Maté et al. (1996)
		Distilled acetylated monoglycerides	
		Glycerol (*p*)	Maté and Krochta (1998)
Pecans (pecan kernels)	Methyl cellulose (MC), hydroxypropyl cellulose (HPC), carboxymethyl cellulose (CMC)	Propylene glycol (PG), sorbitol (*p*)	Baldwin and Wood (2006)
		Lecithin (*s*)	
		α-tocopherol (vitamin A), butylated hydroxyanisole (BHA), butylated hydroxytoluene (BHT) (*ao*)	
Almonds	Hydroxypropyl methylcellulose (HPMC)	Tween 80 (*s*)	Atarés et al. (2011)
		Ascorbic acid, Citric acid, Ginger oil (*ao*)	

p plasticizer, *s* surfactant, *am* antimicrobial, *ao* antioxidant

9.4 Edible Coatings for the Extension of Shelf Life

Other than maintaining organoleptic and other quality attributes, the latest advances in the application of edible coatings have been developed to extend the shelf life of nuts. A number of studies on the oxygen and aroma barrier properties of edible films have been published between 1967 and 2005 presenting varied potential film models (Miller and Krochta 1997). Later studies (Table 9.1) show that suitable film-forming materials include whey protein isolate and cellulosic derivatives which are all good barriers to oxygen (Han and Krochta 2007; Maté and Krochta 1998; Lee and Krochta 2002; Trezza and Krochta 2002; Miller and Krochta 1997), which make them suitable for this use. The same studies state that these materials are good barriers to aroma and oil.

Lin and Krochta (2005) dealt with the adherence of whey protein coatings on peanuts which is vital for shelf-life extension. Peanuts were immersed in aqueous solutions of the surfactants and air dried to modify the peanuts' surface energy which rendered it more compatible with the hydrophilic coating. The addition of surfactants (Span 20) to the whey protein coating solution also improved the coverage of the peanuts in a concentration-dependent fashion Maté et al. (1996) investigated humidity in peanuts and its effect on coatings, and concluded that the factors determining the effectiveness of the coatings were their thickness and the relative humidity, which indicated that the mechanism of protection was due to oxygen barrier properties. The lack of discontinuities of the coatings was critical in improving the shelf life of peanuts.

Protein coatings have been found to be beneficial in that whey protein coatings delayed oxygen uptake and rancidity of dry roasted peanuts when compared to the uncoated controls. Maté and Krochta (1998) found that coated peanuts were definitely more resistant to oxidation than those uncoated and hexanal levels which is a degradation product of linoleic acid and an indicator of lipid oxidation and therefore rancidity was found to be lower (Lee et al. 2002; Lee and Krochta 2002) confirm this fact. This led to the conclusion that coated samples were oxidized significantly slower than the uncoated reference, meaning a shelf-life extension of coated samples.

Other studies such as that of Baldwin and Wood (2006) investigated the application of cellulosic edible coatings on pecan kernels using two plasticizers (propylene glycol and sorbitol), lecithin as surfactant, and several antioxidants (vitamin E, BHA, and BHT). The coatings imparted gloss to the nuts. Determination of degradation products (hexanal) by gas chromatography revealed that hexanal levels were at least twice as low in coated kernels as in uncoated controls, meaning that coated kernels underwent less fat oxidation and were less rancid, which correlated with the sensory analysis, thus further proving the validity of selected coating materials for the preservation of quality and product shelf-life extension.

9.5   Application of Edible Coatings to Almonds

Almonds are one of the most commonly used nuts and are a highly nutritional source of vitamins (B complex) and minerals (Mg, P, K) (Ahmad 2010). They contain high levels of fat (51 % w/w), but have a favorable fatty acid profile (64–82 % oleic acid, 8–28 % linoleic acid, 6–8 % palmitic acid). Here too, the oxidation process of unsaturated fatty acids can produce unpleasant rancid off-flavors, thus shortening the shelf life of almonds under ambient conditions. Almonds with hydroxypropyl methylcellulose (HPMC) formulations containing different antioxidant additives such as ascorbic and citric acids, have been frequently studied. Other materials which may contribute to the organoleptic properties of the nut are potential additives to coating materials. Atarés et al. (2011) added ginger essential oil into HPMC coatings to be applied to almonds, and studied the oxygen permeability of the films in order to characterize the mechanisms that modulated the lipid oxidation protection. Results in Table 9.2 show that acids incorporation produced a significant improvement of the oxygen barrier performance of the films. However, the addition of ginger oil had opposite negative effect on film integrity. This led to the conclusion that whenever materials were added to coating systems they had to be tested rigorously as any material added may affect the coating efficacy.

9.6 Conclusion

Edible films and coatings are a promising alternative to conventional packaging systems that among their functions can be added maintenance of organoleptic and other quality attributes. Many functions of edible packaging are identical to those of plastic films. However, their use would still require an overpackaging, notably for handling and hygiene purposes (Debeaufort et al. 1998).

The use of a coating acting as an oxygen barrier combined with a simpler plastic film acting as a moisture barrier represents an alternative packaging system. The film-forming techniques are critical for the performance of the films (Maté and Krochta 1998) and more work is needed to improve the efficiency of the coatings and achieve longer rancidity delay with thinner films.

Microbial stability will become more and more important as more edible polymers approach commercial availability (Miller and Krochta 1997). And its effect on other properties of the films and the product (oxygen permeability, organoleptic properties, and others) also needs further investigation.

The mechanism of action of each specific compound or mixture should be considered individually to get the best match of the coating formulation and the requirements of the product. Such studies and the exploitation of these results have impact on the quality of nuts in general.

References

Abbey M, Noakes M, Belling GB, Nestel PJ (1994) Partial replacement of saturated fatty acids with almonds or walnuts lowers total plasma cholesterol and low-density-lipoprotein cholesterol. Am J Clin Nutr 59:995–999

Ahmad Z (2010) The uses and properties of almond oil. Complement Ther Clin Pract 16:10–12

Atarés L, Bonilla J, Chiralt A (2010) Characterization of sodium caseinate-based edible films incorporated with cinnamon or ginger essential oils. J Food Eng 100:678–687

Atarés L, Pérez-Masiá R, Chiralt A (2011) The role of some antioxidants in the HPMC films properties and lipid protection in coated toasted almonds. J Food Eng 104:649–656

Ayranci E, Tunc S (2003) A method for the measurement of the oxygen permeability and the development of edible films to reduce the rate of oxidative reactions in fresh foods. Food Chem 80:423–431

Baldwin EA, Wood B (2006) Use of edible coating to preserve pecans at room temperature. HortScience 41(1):188–192

Basaran P, Basaran-Akgul N, Oksuz L (2008) Elimination of Aspergillus parasiticus from nut surface with low pressure cold plasma (LPCP) treatment. Food Microbiol 25:626–632

Biquet B, Labuza TP (1988) Evaluation of the moisture permeability of chocolate films as an edible moisture barrier. J Food Sci 53(4):989

Chidanandaiah RCK, Sanyal MK (2009) Effect of sodium alginate coating with preservatives on the quality of meat patties during refrigerated (4±1 °C) storage. J Muscle Foods 20:275–292

Contreras-Medellin R, Labuza TP (1981) Prediction of moisture protection requirements for foods. Cereal Food World 26(7):335

Cosler HB (1957) Methods of producing zein-coated confectionery. US Patent 2,791,509

Debeaufort F, Quezada-Gallo JA, Voilley A (1998) Edible films and coatings: tomorrow's packagings: a review. Crit Rev Food Sci 38(4):299–313

Duan J, Cherian G, Zhao Y (2010) Quality enhancement in fresh and frozen lingcod (Ophiodon elongates) fillets by employment of fish oil incorporated chitosan coatings. Food Chem 119:524–532

Frankel EN (1996) Antioxidants in lipid foods and their impact on food quality. Food Chem 57(1):51–55

Gómez-Estaca J, Montero P, Giménez B, Gómez-Guillén MC (2007) Effect of functional edible films and high pressure processing on microbial and oxidative spoilage in cold-smoked sardine (Sardina pilchardus). Food Chem 105:511–520

Han JH, Krochta JM (2007) Physical properties of whey protein coating solutions and films containing antioxidants. J Food Sci 72(5):308–314

Hong SI, Krochta JM (2006) Oxygen barrier performance of whey-protein-coated plastic films as affected by temperature, relative humidity, base film and protein type. J Food Eng 77:739–745

Kurlandsky SB, Stote KS (2006) Cardioprotective effects of chocolate and almond consumption in healthy women. Nutr Res 26:509–516

Labuza TP (1971) Kinetics of lipid oxidation in foods. CRC Crit Rev Food Technol 2:355–405

Lee SY, Krochta JM (2002) Accelerated shelf life testing of whey-protein-coated peanuts analyzed by static headspace gas chromatography. J Agric Food Chem 50:2022–2028

Lee SY, Trezza TA, Guinard JX, Krochta JM (2002) Whey-protein-coated peanuts assessed by sensory evaluation and static headspace gas chromatography. J Food Sci 67(3):1212–1218

Lin SD, Krochta JM (2005) Whey protein coating efficiency on surfactant-modified hydrophobic surfaces. J Agric Food Chem 53:5018–5023

Lopez-Caballero ME, Gómez-Guillén MC, Pérez Mateos M, Montero P (2005) A chitosan-gelatin blend as a coating for fish patties. Food Hydrocoll 19:303–311

Luttfullah G, Hussain A (2011) Studies on contamination level of aflatoxins in some dried fruits and nuts of Pakistan. Food Control 22:426–429

Maté JI, Krochta JM (1998) Oxygen uptake model for uncoated and coated peanuts. J Food Eng 35:299–312

Maté JI, Frankel EN, Krochta JM (1996) Whey protein isolate edible coatings: effect on the rancidity process of dry roasted peanuts. J Agric Food Chem 44:1736–1740

Miller KS, Krochta JM (1997) Oxygen and aroma barrier properties of edible films: a review. Trends Food Sci Technology 8:228–237

Nussinovitch A (2009) Biopolymer films and composite coatings. In: Kasapis S, Norton IT, Ubbink JB (eds) Modern biopolymer science. Elsevier, San Diego

Ou S, Wang Y, Tang S, Huang C, Jackson M (2005) Role of ferulic acid in preparing edible films from soy protein isolate. J Food Eng 70:205–210

Oussalah M, Caillet S, Salmiéri S, Saucier L, Lacroix M (2004) Antimicrobial and antioxidant effects of milk protein-based film containing essential oils for the preservation of whole beef muscle. J Agric Food Chem 52:5598–5605

Pérez-Gago MB, Serra M, del Río MA (2006) Color change of fresh-cut apples coated with whey protein concentrate-based edible coatings. Postharvest Biol Technol 39:84–92

Ponce A, Roura SI, del Valle CE, Moreira MR (2008) Antimicrobial and antioxidant activities of edible coatings enriched with natural plant extracts: in vitro and in vivo studies. Postharvest Biol Technol 49:294–300

Swenson HA, Miers JC, Schultz TH, Owens HS (1953) Pectinate and pectate coatings. Applications to nuts and fruit products. Food Technol 7:232–235

Tapia MS, Rojas-Graü MA, Carmona A, Rodríguez FJ, Soliva-Fortuny R, Martin-Belloso O (2008) Use of alginate and gellan based coatings for improving barrier, texture and nutritional properties of fresh-cut papaya. Food Hydrocoll 22:1493–1503

Tharanathan RN (2003) Biodegradable films and composite coatings: past, present and future. Trends Food Sci Technol 14:71–78

Trezza TA, Krochta JM (2002) Application of edible protein coatings to nut and nut-containing food products. In: Gennadios A (ed) Protein-based films and coatings. CRC, Boca Raton

Chapter 10
Kaanga Wai: Development of a Modern Preservation Process for a Traditional Maori Fermented Food

John D. Brooks, Michelle Lucke-Hutton, and Nick Roskruge

Contents

10.1 Introduction

Kaanga wai (or kaanga kopiro) is a traditional fermented food produced by the Maori of New Zealand from maize (*Zea mays*). It has been prepared by Maori since the early 1800s (Asmundson et al. 1990). The traditional process involved putting whole cobs into woven flax bags and immersing them either in a stream of running water or in slow-moving swamp water (Whyte et al. 2001). The bags were kept submerged for 2–3 months, after which the maize was judged ready by squeezing

J.D. Brooks (✉) • M. Lucke-Hutton • N. Roskruge
Faculty of Health and Environmental Sciences, School of Applied Sciences,
Auckland University of Technology, 24 St. Paul Street, Auckland, New Zealand
e-mail: foodmicrobiologist.007@gmail.com

© Springer Science+Business Media New York 2016
A. McElhatton, M.M. El Idrissi (eds.), *Modernization of Traditional Food Processes and Products*, Integrating Food Science and Engineering Knowledge Into the Food Chain 11, DOI 10.1007/978-1-4899-7671-0_10

the kernels to check that they had softened. The cobs were removed from the bags and kernels were scraped from the cobs before grinding using a stone mortar (Asmundson et al. 1990). The distinctive flavour was highly prized by some Maori and the product was consumed as a porridge, often with cream and sugar (Asmundson et al. 1990), but it was also boiled and mixed with manuka ash in a product known as Kaanga pungarehu (Roskruge 2007), or dried and used like flour in cake and bread making.

Modern production of Kaanga wai was almost non-existent until a revival in the last 20 years or so—pollution of streams, theft and the loss of suitable cultivars were significant factors, but the influence of western missionaries on Maori to give up their "unwholesome" and "uncivilised ways" and changes in the attitudes of young Maori also resulted in a decline in the number of people with the knowledge and experience to produce Kaanga wai. There is also very little published material on production and qualities of Kaanga wai.

The modern, non-commercial fermentation process differs little from the traditional method, but there has been some evolution to use more modern materials, such as hessian sacks, sometimes with muslin or cotton flour bag lining, instead of the flax bags. In some cases, the traditional whole cob process has been replaced by stripping the kernels from the cobs before the fermentation. Also, some practitioners are using drums with water reticulation to replace the need to place the cobs in natural sites.

Programmes have been founded to aid the re-establishment of the traditional varieties of corn and maize (generally Indian corn) for fermentation; many modern cultivars contain too much sugar to be suitable for Kaanga wai production (personal communication, Taanehopuwai Trust 2003).

10.2   The Traditional Fermentation Process

The best maize variety to use for the fermentation is an old cultivar referred to simply as "Kaanga" (Taanehopuwai Trust 2003), which is nuttier in flavour and is a much darker colour than sweet corn. These traditional varieties are generally referred to in literature as "Indian corn" and are maize cultivars which have open pollinated over generations. One old cultivar with very white kernels, found in the Waikato region of New Zealand, was known as Niho hoiho (horse's teeth) because of the size of the kernels and this was the old preference.

The two methods of preparation currently used are whole cob, and fermentation of kernels stripped from the cob (Figs. 10.1 and 10.2). In both methods, the cobs are left on the plant until the kernels have dried and hardened, judged on the plant once the cobs hang. This translates to a moisture level in the kernels of 12–14 % at harvest (Roskruge 2007). The outer leaves of the cob are left on and the whole cob is placed in a jute or hessian bag. The bag is completely immersed in running water in a pool in a creek and left for 2–3 months to ferment. At intervals the corn is checked by squeezing between finger and thumb—if the kernels are soft, then the corn is ready; if the kernels are still hard, the bag is put back into the water. If the kernel method

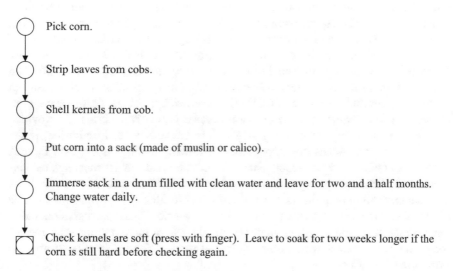

Pick corn, leave leaves on cob.

Put corn into a sack (made of hessian or jute).

Immerse sack containing corn in running water and leave for two months. A pool in a creek is acceptable.

Check kernels are soft (press with finger). Leave to soak for two weeks longer if the corn is still hard before checking again.

Fig 10.1 Flow diagram for the whole Cob method

Pick corn.

Strip leaves from cobs.

Shell kernels from cob.

Put corn into a sack (made of muslin or calico).

Immerse sack in a drum filled with clean water and leave for two and a half months. Change water daily.

Check kernels are soft (press with finger). Leave to soak for two weeks longer if the corn is still hard before checking again.

Fig. 10.2 Flow diagram for the Kernel method

is being used, the kernels are removed from the cob and placed in a muslin or calico bag, which is then placed in a drum of clean water. The water must be changed every day to simulate running fresh water. Again, the kernels are left for 10 weeks and checked for softness at intervals.

10.3   Microbiology and Safety

Since the Kaanga wai fermentation is uncontrolled and no starter culture is used, it is important to understand the course of the fermentation and the microbes to ensure that all appropriate steps are taken to minimise risks to the consumer.

Both aerobes and anaerobes can be isolated, together with typical fermentation products, such as lactic acid and volatile fatty acids (VFA) (Asmundson et al. 1990). Lactic acid is produced between 2 and 20 days, but is slowly replaced by VFA. At 1.5–2.5 months propionic acid appears, but *n*-butyric acid, which is probably what gives the product its characteristic smell, predominates by the time the process is considered complete.

As in most natural fermentations, there is a succession of bacterial types observed. Initially lactic acid bacteria predominate—strains of *Lactococcus* and *Leuconostoc* dominate the early fermentation and during this time the lactic acid concentration rises rapidly. *Lactococcus* disappears from the fermentation after 23 days, but *Leuconostoc* survives for more than 67 days (Asmundson et al. 1990).

A major concern with natural fermentations is that the pH may not fall sufficiently to inhibit *Clostridium botulinum*, a spore-forming strict anaerobe that produces botulin, a neurotoxin that is extremely toxic even at low concentrations (Byrne et al. 1998). Asmundson et al. (1990) isolated a butyric acid producing *Clostridium* species from their samples, indicating that *C. botulinum* could potentially be present in the product. However, it appears that *C. botulinum* does not compete well with large numbers of other microorganisms (Montville 2008) and toxin containing foods are generally devoid of other types of bacteria because of thermal processing (Jay et al. 2005). However, Whyte et al. (2001) have concluded that the Kaanga wai preservation process is relatively low risk, provided that there is an adequate cooking step before consumption. In addition, if the fermentation process has failed, potentially allowing pathogens to grow, there is reported to be an obvious strong, sharp, bitter taste that would warn against significant accidental consumption. It should be noted that botulin is so toxic that a taste sample might contain sufficient toxin to cause illness. However, the toxin is heat labile, so tasting a cooked sample would be safe. These workers concluded that normal safe food handling practices and the use of uncontaminated water for the production process would be sufficient to prevent food poisoning from consumption of this food.

10.4   Packaging

Traditionally, the Kaanga wai was consumed as soon as it was ready to eat. However, the Taanehopuwai Trust (a member of the Tahuri Whenua collective, www.tahuri-whenua.org.nz) wanted to develop a shelf-stable product that could be stored and prepared easily. A retortable pouch was the obvious choice. These packages have been described by Downing as consisting of a flexible pouch-shaped container generally made of a three-ply laminate consisting of polyester film, aluminium foil and polypropylene film to give "superior barrier properties for a long shelf life, seal integrity, toughness and puncture resistance" (Downing 1996). The retortable pouch is currently used for packaging a variety of food products, including meats, sauces, soups, fruits, vegetables and pet food.

10.5   Thermal Processing

Foods packed and stored in hermetically sealed containers must be heat processed to ensure safety, particularly if the product pH is above 4.5 since these conditions may allow the growth and toxin production of *C. botulinum* (Weddig et al. 2007). The thermal process is designed using the known heat resistance of *C. botulinum* spores and the measured heat penetration into the slowest heating point (SHP) of the product. It is critical to determine the SHP accurately to ensure that the entire product receives an adequate thermal process. To ensure that worst-case conditions were accounted for in the development of the thermal process, an assumption was made that Kaanga wai would be classified as a low acid food, since its pH may be greater than 4.5, so by regulation, it must be given a thermal process that will reduce the number of *C. botulinum* spores present by a factor of 10^{12}. This is referred to as a 12-D process (Weddig et al. 2007)

10.6   Determination of the Thermal Process

Retortable pouches have a complex geometry, particularly if there is a gusset at one end. To measure the heat penetration and determine the SHP, thermocouples were inserted into the pouches through the base and held in place by plastic pillars (Figs. 10.3 and 10.4).

The pillars ensured that the thermocouples remained in the desired position within the pouch and were made of plastic, rather than metal, to minimise the heat sink effect. The thermocouple was secured in the base of the pouch, using a rubber washer on the outside of the pouch and a brass washer and nut on the inside, thus effectively sealing the pouch so that it did not leak during retorting. The pillars were secured through the sides of the pouch using rubber o-rings and cap screws. Pillars were placed at three different positions within the pouches to enable the SHP to be determined.

The quantities of Kaanga wai presently available are relatively small. The pouches were therefore filled with 540 g of Cream Style Corn (Wattie's, Hasings, New Zealand) and vacuum-sealed at 350 mbar. This product is similar to the fermented Kaanga wai and provided a convenient model system, though it should be noted that Cream Style Corn has already been processed and the starch is therefore gelatinised, altering the thermal conductivity of the product. An automatically controlled retort (steam pressure vessel) with an Allen-Bradley supervisory control and data acquisition system (Rockwell Automation, Milwaukee, Wisconsin) was used to process sealed pouches at 115 °C for 90 min (see Fig. 10.5). The temperature profiles from the thermocouples were recorded and the real time value of F_0 was automatically calculated using Simpson's rule (Timings and Twigg 2001).

Fig. 10.3 Thermocouple held in place in retortable pouch

Fig. 10.4 Pouch assembled ready for retorting

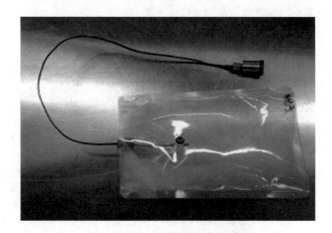

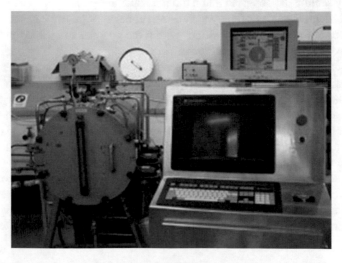

Fig. 10.5 Automatically controlled and monitored retort (steam pressure vessel)

From the temperature–time profile, the lethal rate can be determined from the equation:

$$L = 10^{((\theta - \theta_{ref})/z)} \tag{10.1}$$

where L=lethal rate, θ=product temperature (°C), θ_{ref}=reference temperature (°C), and z=the temperature range over which the D-value decreases by a factor of 10 (°C).

From the lethal rate, the lethality (or equivalent time) of the process can be calculated, using the following integration:

$$F = \int_0^t L dt = \int_0^t \left(10^{((\theta - \theta_{ref})/z)}\right) dt \tag{10.2}$$

where F=lethality (F_0=lethality using a reference temperature of 121.11 °C and z=10 °C). L=lethal rate. So, for a $12D$ process:

$$F_0 = D_r \left(\log C_0 - \log C\right) = 12 D_r,$$

where F_0=integrated lethal value of heating using a reference temperature of 121.11 °C and z=10 °C. D_r=value of D determined at 121.1 °C, C_0=initial concentration of spores, and C=final concentration of spores.

A local processor of canned corn recommended an F_0 of 9 min, which, taking D_r as 0.21 min and z as 10 °C would result in a 42 decimal reduction in spores, ensuring the safety of the finished product. The SHP was found to lie between 60 and 80 mm from the base of the pouch and the final process design was a processing time of 100 min at 115 °C.

10.7 Thermal Processing of Kaanga Wai

The Kaanga wai was prepared for processing according to the following flow diagram (Fig. 10.6).

The appearance of the minced product was quite different from the traditional product, which is made by pounding the whole kernels with the bottom of a strong glass bottle. Mincing of the kernels resulted in a smaller particle size, with more of the kernel contents released, increasing the starch content of the liquid fraction and giving a drier appearance.

Traditional preparation of the Kaanga wai involves mixing the mashed kernels with water in the ratio 1:4 (Kaanga wai:water). For this reason, various ratios of water to corn were processed and compared. Where higher ratios of water were used, there was a tendency for the product to separate before thermal processing. This problem was resolved by pre-gelatinising the product in water at a ratio of Kaanga wai:water of 1:2 before filling into pouches. The finished product was homogeneous, though slightly too thick. The final process could therefore include a pre-gelatinisation step, or the pouches could be processed in an agitating retort.

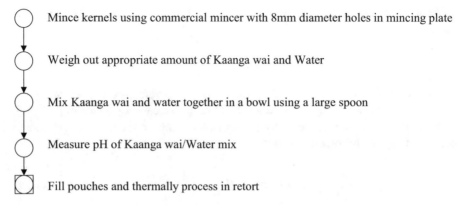

○ Mince kernels using commercial mincer with 8mm diameter holes in mincing plate

○ Weigh out appropriate amount of Kaanga wai and Water

○ Mix Kaanga wai and water together in a bowl using a large spoon

○ Measure pH of Kaanga wai/Water mix

◻ Fill pouches and thermally process in retort

Fig. 10.6 Kaanga wai preparation

The large amount of water required to produce a product of the required consistency suggests that there is a large amount of starch still present in the fermented corn and implies that little starch is hydrolysed during the fermentation.

10.8   Comparison of Traditional Kaanga Wai with Model Cream Style Corn during Thermal Processing

When the lethality curves for commercially minced Kaanga wai were compared with the Cream Style Corn model, it was found that heat transfer occurred much more rapidly into the Kaanga wai. This is almost certainly because the Cream Style Corn had already been gelatinised by its earlier processing and was thus more viscous, whereas the Kaanga wai thickened only during the heating process. The use of the model is therefore not valid, though the product would be over-processed, rather than potentially dangerously underprocessed.

10.8.1   Sensory Evaluation

Preliminary sensory evaluation of the new product was undertaken with a consumer panel drawn from local Maori familiar with the traditional product. This in itself presented difficulties: there has been little production of Kaanga wai for many years and finding sufficient Maori who knew the product was challenging. The retorted product was compared with a sample of the traditional Kaanga wai, which was prepared by adding two cups of the fermented corn to four cups of boiling water and simmering for 1 h. The retorted product was slightly too thick, so a small amount of water was mixed into the product before it was heated to boiling point. Both products were kept at 65 °C in a Bain-Marie prior to being served to the panellists.

The products were compared with a difference test designed to detect whether there is an overall difference between the products rather than a difference in a specific characteristic. The test used was a two-out-of-five test. The number of panellists required to take part in this panel in order to give statistically significant results is between 10 and 15 people.

A nine-point hedonic scale test was used for acceptance testing. Panellists were asked to rate two samples presented (one of each product). The number of panellists required to take part in this panel in order to give statistically significant results is between 60 and 100 people.

The numbers of panellists available fell below these minima, so the results are not statistically significant, but do give an indication of consumer response to the product. In the difference test, only one out of seven panellists correctly identified the two samples that were the same. There was no difference in acceptance between the two samples; both had a mean acceptance rating of "Like slightly". This may, however, be misleading, as panellists generally fell into two groups—either they liked the product very much (5/7) or they disliked the product (2/7).

10.8.2   Shelf Life Testing

It is impossible to conduct accelerated shelf life testing of the retorted Kaanga wai as there are no substantially similar products on the market with which to compare results. Shelf life testing must therefore take place over a full year, with appropriate tests being conducted every month. Suitable trials would involve two sets of samples, one set being frozen to provide a reference and the other stored at constant temperature and humidity (20 °C and 75 % RH). These constraints are critical, as moisture and gas exchange with the product through the laminate of the pouch is possible. So far, such trials have not been conducted. Ideally, appearance, texture and flavour of the product should be tested by a trained panel, using difference and acceptance tests. If resources are limited, a larger consumer panel could be used. This presents some problems for Tahuri Whenua, as in order to ensure accuracy and repeatability of results; at least 40 consumers will be needed each month to test the samples. Other analyses that should be made at the monthly intervals are colour, viscosity and texture measurements.

If the sensory panel results are shown to be correlated with instrumental measurements, then quality standards can be developed that will allow instrumental tests to be used to determine acceptability of the product to consumers.

10.9   Legal Issues

Because of the implications of the New Zealand Food Act (1981) and the unique position of Maori as partners in the Treaty of Waitangi, there are a number of issues that may have to be resolved if this process is developed further, especially those relating to intellectual property.

The product is partially processed at an unlicensed premise, i.e. the corn is fermented in a stream. This will become important if the product is sold to consumers outside the Iwi (tribe), as the product would then be considered to be commercial and would therefore be required to meet the Food (Safety) Regulations 2002 (New Zealand Government 2002). This specific issue might be addressed through the application of HACCP principles and effectively regarding the fermentation phase as "harvesting". Perhaps the aspect of most concern during the fermentation phase is the potential for chemical contamination of the product from farm run-off and industrial discharge. Proper application of HACCP would require that the water be tested for presence of industrial and agrichemicals on a regular basis.

Care must also be exercised in the handling of the raw fermented corn. The fermentation occurs under essentially uncontrolled conditions, so whatever microorganisms present in the water and on the corn may take part in the fermentation process. Some of these microorganisms may be harmful to the workers or may produce toxins in the raw material.

During the development of the process described above, the pH of the fermented product was found to lie between 3.67 and 3.75. Further testing will be required on many batches of Kaanga wai to determine whether it is a low acid food (i.e. has a pH greater than 4.5). If it is found that the pH is occasionally greater than 4.5, then the product will either require acidification or be processed as a low acid food. In either case, a scheduled thermal process will have to be filed and the product will have to be processed according to the schedule at an approved food processing premises, in accordance with Regulation 14 of the Food Safety Regulations 2002 (NZFSA 2011).

The second issue is that of intellectual property. The fermentation process is a traditional method employed by the New Zealand Maori (and incidentally, in the production of "tocos" by the Ancash Indians of Peru) and should probably be regarded as public property. However, under the Treaty of Waitangi, the rights of Maori must be carefully considered in relation to traditional plants, medicines and foods. A problem could arise if any individual or group were to try to patent the process or to develop the process commercially and then patent it. Ownership and use of traditional knowledge is a major issue for both Maori and Pakeha (non-Maori) and has been the subject of the WAI262 claim to the Waitangi Tribunal. The report of the Tribunal was released in July, 2011 and will influence future law and policy affecting Māori culture and identity, native flora and fauna (Waitangi Tribunal 2011).

Further investigation of all these issues will be essential if the process is to be fully commercialised and the product sold to the general public.

10.10   The Future

The best option for the Tahuri Whenua at this stage of development of the product would be to prepare the raw material to the processing stage and then contract a suitable food manufacturer to package the product and apply a heat treatment, possibly in accordance with a scheduled process.

References

Asmundson RV, Boland MJ, Moore DWG, Davis W, Winiata W (1990) Kaanga Wai, a New Zealand Maori corn fermentation. In: Yu P (ed) Fermentation technologies: industrial applications. Elsevier, London

Byrne MP, Smith TJ, Montgomery VA, Smith LA (1998) Purification, potency, and efficacy of the Botulinum Neurotoxin type A binding domain from Pichia pastoris as a recombinant vaccine candidate. Infect Immun 66:4817–4822

Downing DL (1996) A complete course in canning and related processes. Book II: microbiology, packaging, HACCP and ingredients. CTI, Baltimore

Jay JM, Losesner MJ, Golden DA (2005) Modern food microbiology. Springer, New York

Montville TJ (2008) Food microbiology. McGraw-Hill, New York. http://www.accessscience.com. Accessed 13 Sept 2011

New Zealand Food Safety Authority (2011) Food Safety Regulations 2002

New Zealand Government (1981) Food Act 1981

New Zealand Government (2002) Food (Safety) Regulations 2002 (SR 2002/396). New Zealand Government, Wellington

Roskruge N (2007) Hokia ki te whenua, Ph.D. thesis, Massey University, Palmerston North

Taanehopuwai Trust (2003) Kaanga Wai production

Timings R, Twigg P (2001) Dictionary of engineering terms. Butterworth-Heinemann, Boston

Waitangi Tribunal (2011) Ko Aotearoa tēnei: a report into claims concerning New Zealand law and policy affecting Māori culture and identity. Te taumata tuatahi. Government of New Zealand

Weddig LM, Balestrini CG, Shafer BD (2007) Canned foods: principles of thermal process control, acidification and container closure evaluation. Science Education Foundation, Washington, DC

Whyte R, Hudson JA, Hasell S, Gray M, O'Reilly R (2001) Traditional Maori food preparation methods and food safety. Int J Food Microbiol 69:183–190

Chapter 11
Thai Fish Sauce: A Traditional Fermented Sauce

Wunwiboon Garnjanagoonchorn

Contents

11.1   Introduction

Thai fermented fish product—fish sauce is known locally as "Nam Pla." Fish fermentation is widely practiced in Thailand and fish sauce is widely consumed in the country and is also the most important exported fish fermented product.

The process of fermentation for fish sauce starts when the raw material and salt are left to stand. During the fermentation process of fish sauce, some organic substances break down into smaller molecules which contribute to the typical odor, flavor, and color to the products. Similar products are also processed and consumed in other countries in the Southeast Asia, i.e., Malaysia, Laos, Vietnam, Cambodia, Myanmar, Indonesia, and the Philippines.

W. Garnjanagoonchorn (✉)
Department of Food Science and Technology, Faculty of Agro-Industry,
Kasetsart University, Bangkok, Thailand
e-mail: wunwiboon.g@ku.ac.th

© Springer Science+Business Media New York 2016
A. McElhatton, M.M. El Idrissi (eds.), *Modernization of Traditional Food Processes and Products*, Integrating Food Science and Engineering Knowledge Into the Food Chain 11, DOI 10.1007/978-1-4899-7671-0_11

11.2   Fish Sauce

Fish sauce or "Nam Pla" in Thai is the clear aqueous product of prolonged salting fish fermentation. It is made from either freshwater or saltwater fish. Anchovies (small saltwater fish) are traditionally used in good quality fish sauce; however, fish and parts of fish can be used in fish sauce fermentation (Phithakpol et al. 1995; Lopetcharat 1999) Natural fish sauce requires 9–12 months fermentation. Fish sauce is used as a flavoring ingredient in Thai cooking as well as other Southeast Asian cooking; it is called differently such as Nuoc-nam (Vietnam), Gua-ca (Myanmar), Kecap ikan (Indonesia), Patis (Philippines), and Nam-pa (Laos). Good quality fish sauce imparts good aroma and flavor, contains essential amino acids, is rich in vitamin B especially vitamin B_{12}, and also supplies minerals that include calcium, phosphorous, iodine, and iron (Areekul et al. 1974; Garby and Areekul 1974; Suwanik 1977). Thai fish sauces (Fig. 11.1) are produced for local consumption and for export to nearby countries and others such as the European Union, the USA, Canada, Japan, Australia, and New Zealand.

In Thailand, fish sauce is classified as "standardized food" where quality standards are defined by regulation authorized by Food and Drug Administration, under the Ministry of Public Health (FDA 2000). Genuine fish sauce shall be clear and free of sediment, has color, odor, and flavor inherent of that specific characteristics of genuine fish sauce, if only sodium chloride salt is used it shall be not less than 200 g in 1 L of fish sauce, total nitrogen not less than 9 g/L of fish sauce, amino acid nitrogen not less than 40 % and not more than 60 % of total nitrogen, and ratio of glutamic acid to total nitrogen not less than 0.4 and not more than 0.6 (Ministry of Public Health 2000). Histamine, a toxic biogenic amine derived from enzymatic decarboxylation of amino acid histidine, has a maximum permitted level of 400 mg/ kg food (Codex Std 302-2011). Using spectrofluorometry technique, Muangthai and Nakthong (2014) analyzed ten fish sauces bought from supermarkets in Bangkok and found histamine content in the range of 7.5–15.11 mg/kg fish sauce. The low level of histamine indicates the good quality of fish raw material used in fish sauce manufacturing.

Fig. 11.1 Example of the Thai commercial fish sauce

11.3   Manufacturing Process

Natural fish sauce is made from the very fresh whole fish mostly small fish, often salting on board after catch. Fish are mixed with salt at a variable ratio from 4:1 to 1:1 (w/w); the ratio of fish to salt depends on size of fish and traditional recipe of each manufacturer. Either solar (sea) salt or rock salt can be used in the process. The fish and salt mixture is placed in large earthen or concrete containers lined with a layer of salt on the bottom, and topped with a layer of salt. Figure 11.2a shows the fermentation of fish sauce in concrete containers exposed to sunshine during the day. Figure 11.2b, c showed the fermentation in glass tanks that permitted observation of the changes of the mixture during fermentation. During the first few months, a woven bamboo mat with heavy rocks is placed on the top layer to prevent the fish in the mixture from floating during fermentation. Approximately after a period of 6 months, the light brown colored liquid is removed from the fish mixture through an outlet at the bottom of the container that facilitates separation of the sediment from the liquid. This liquid is then allowed to ferment in the sun until good aroma and flavors are developed. This top grade fish sauce is corrected to meet quality standards and is then ready for bottling. The high quality fish sauce requires about 9–12 months fermentation; therefore, it is quite challenging for researchers to find means of reducing fermentation time and still produce good aroma and flavors fish sauce. The remaining sediment from the fermentation process is mixed with brine solution, and retained for a period of 2–3 months to extract fish flavors; it is then filtered and bottled as second- and third-grade fish sauce. In commercial fish sauce, sugar, caramel color, and monosodium glutamate may be added to adjust color and flavor.

11.4   Research and Development of Thai Fish Sauce

Thai fish sauce has been traditionally processed in households and small-scale industries where research and development have been carried out by processors and kept secret within the families. Over the past 40 years, many scientific studies had been carried on Thai fish sauce and published for public information. The topics of these studies involved microbiology and chemistry of fish sauce, also methods to improve fish sauce processing and quality.

Fig. 11.2 Fish sauce fermentation, (**a**) fermentation tank, (**b, c**) the changes of fish and salt mixture during fermentation

11.4.1 Microbiology and Chemistry of Fish Sauce

During the 1–2 month of fish sauce fermentation, the concentration of soluble nitrogen compounds in fish and salts mixture increases which involve endogenous enzyme in fish muscle and viscera. The activity of trypsin-like proteinase has been reported by Gildberg and Shi (1994) and Sirighan et al. (2006), also cathepsin (Lopetcharat and Park 2002) during the first period of fermentation. However, high salt concentration (15–20 %) and low pH (5.5) condition do not favor these proteinase activities.

The changes of microorganism during the beginning of fish sauce fermentation were studied by Thongthai and Siriwongpairat (1978) who demonstrated that after the first month of fermentation the number of aerobic bacteria that tolerate 5–10 % salt that present in a small number will decrease and the aerobic halo-tolerant bacteria will increase from 10^7 to 7×10^8 cells/mL after 3 weeks of fermentation. More specific bacteria that involved in the first stage of fermentation have been identified by Saengjindawong and Winitnuntarat (1984) who found microorganisms to be *Micrococcus roseus, M.varians, Pediococcus cerevisiae, P. halophilus, Bacillus pumilus, B. megaterium, B. firmus, B. alvei, and B. laterosporus*. Halophilic bacteria that showed high proteinase activity are also found during fish sauce fermentation (Thongthai et al. 1992; Chaiyanan et al. 1999; Hiraga et al. 2005). In the late stage of fermentation, many groups of bacteria have been isolated by researchers. Saisithi et al. (1966) reported that *Staphylococcus* sp., *Bacillus* sp., coryneforms, *Streptococcus* sp., and *Micrococcus* sp., isolated from Thai fish sauce after 9 months fermentation were responsible for producing volatile acids that impart good odor in Thai fish sauce. Among these isolates, *Staphylococcus* sp. produced the highest volume of volatile acids. Saisithi (1967) pointed out that five volatile organic acids namely formic, acetic, propionic, isobutyric, and one unknown acid provide Thai fish sauce aroma, and water soluble nitrogen compounds, histidine, proline, and glutamic acid provide fish sauce flavor. There were 20–22 amino acids identified in the first few months of fish sauce fermentation and reduced to 13 amino acids in the final stage of fermentation. This is due to the involvement of amino acids in nonenzymatic browning reactions which resulted in an amber color development in fish sauce (Saisithi 1967, 1968; Raksakulthai and Haard 1992). Most studies (Saisithi et al. 1966; Liptasiri 1975; Suntinanalert 1979; Chaiyanan and Chaiyanan 1983; Thongthai et al. 1992; Chaiyanan 2000) showed that halophilic and halo-tolerant bacteria were involved through the whole process of fish sauce fermentation. Liptasiri (1975) isolated seven halo-tolerant bacteria namely *Bacillus* sp., *Staphylococcus* sp., *Micrococcus* sp., coryneforms, *Streptococcus* sp., *Lactobacillus* sp., and *Sarcina* sp. from Thai fish sauce. She pointed out that the first four species provide protease enzyme during fermentation where the product of enzyme reaction may involve in fish sauce flavors and color development. These groups of bacteria were similar to the reports of Saisithi et al. (1966) and Suntinanalert (1979). Parts of bacteria come from salts used in fermentation, as Suntinanalert (1979) found *Halobacterium* sp., and *Halococcus* sp., in most solar salt samples sampling from different suppliers in Thailand; however, *Bacillus* sp., *Micrococcus* sp., and

Staphylococcus sp. were found in some of the sea salt samples. She also investigated bacteria in rock salt samples in Thailand and found *Micrococcus* sp., *Bacillus* sp., and coryneforms. The high salt content (more than 20 %) in fish and salt mixture during fish sauce fermentation help eliminating spoilage bacteria and promoting growth of halophilic and halo-tolerant bacteria. *Pediococcus halophilus* is important in providing good odor in Thai fish sauce (Chaiyanan and Chaiyanan 1983). Saisithi (1987) recommended that the optimal salt concentration for fish sauce fermentation is 20 % by weight. Samittasiri (1986) isolated halophilic bacteria from fish sauce and found that *Halobacterium salinarium* is the bacteria that provide good odor of fish sauce. Later, Klomklang (1995) isolated bacteria that can produce acid, enzyme protease, lipase and amylase from fish sauce collected from different fermentation time; among these isolates she identified and reported that *Staphylococcus saprophyticus* 0113 can hydrolyze lipids and provide acid; *Bacillus pantothenticus* 0118 can hydrolyze starch and protein; *Halobacterium salinarium* 0509 was able to grow in the presence of 25–30 % sodium chloride; and an unknown Gram-negative rod-shaped bacteria 0406 can hydrolyze starch. Proteolytic bacteria, which can grow well in a medium containing high salt, have been isolated from fish sauce (Chaiyanan and Chaychotcharoen 1987; Chaiyanan 2000). Saisithi (1994) concluded in his book that excess salt addition is costly and will slow down the fermentation rate. Salt could retard protein digestion if its concentration is higher than 15 %. Most traditional fermentation took 9–12 months or more; the long fermentation time may be the result of high salt concentration which retards fish muscle digestion at the beginning of fermentation and also slows down bacterial activities.

Although halophilic bacteria that involved in Thai fish sauce fermentation can be isolated, commercially provided pure culture that can help reducing fermentation time and providing good fish sauce odor and flavor has not been successfully accomplished.

11.4.2   Methods to Improve Fish Sauce Processing and Its Quality

Fish sauce is the product of fish hydrolysis where the solid fish is liquefied; this involves the activities of bacteria and fish enzymes under high salt and micro-aerobic conditions. Among the different methods attempting to reduce fish sauce fermentation time that have been investigated by Thai researchers, the addition of enzymes, either crude or pure, obtained from plants, animals, and bacteria, to fish and salts mixture has been favored (Suwunnasart 1966; Mittranond 1983; Poosaran 1986b; Raksakulthai et al. 1986). Suwunnasart (1966) compared the effect of commercial protease obtained from microorganism (*B. amyloliquefaciens*) and papain. Crude enzyme at concentration of 0.4 % of fish weight was added to minced fish and salt liquid mixture and left to ferment at 37 °C for 68 days. The quality of the enzyme-added fish sauce was reported to be better than the non-enzyme-added sample, where papain-added fish sauce showed the highest amino acid nitrogen content. Gongtip (1965) added bromelain during fermentation and also obtained a good quality fish sauce with shorter fermentation time. Mittranond (1983) produced fish sauce by adding proteinase T (commercial enzyme) and crude pyloric enzyme (prepared from saltwater fish) to hydrolyze fish protein during fermentation. The resulting fish sauce exhibited unsatisfied quality due to the lack of good fish sauce odor. However, the addition of enzyme plus *Halobacterium* (isolated from fish sauce sample) gave good odor liquid fish sauce (Mittranond 1983). It has been shown that addition of crude or pure enzyme can help accelerate fermentation rates; however, extra process time would be needed to help promote the production of volatile compounds that are responsible for good fish sauce odor.

The reduction of fermentation time is also carried by means of the addition of koji (the common name of the fungus *Aspergillus oryzae*). Jongsereejit (1990) screened 12 strains of *Aspergillus oryzae* isolated from the soy sauce koji and sake of Japan, the soy sauce koji of Thailand, and the fish sauce koji of Philippines for protease and amylase activities. Only *Aspergillus oryzae* F and *Aspergillus oryzae* W 215 have the highest protease and amylase activities. In this study, 12.5 % koji from selected strains and 5 % salt were added to anchovy fish and left to ferment for 1 day at 50 °C then added salt to 25 % and incubated at 40 °C for 7 days. The resultant hydrolysate had a good odor, flavor, and reddish brown color fish sauce with 28.93 % sodium chloride and amino acid nitrogen of 11.84 mg/mL. The incorporation of koji appeared to accelerate protein hydrolysis, color development as well as flavor and aroma development. It should also be emphasize that protease activity is inhibited by the increase in salt concentration. In 1995, Klomklang studied the fermentation of fish sauce by using halophilic bacteria with koji. In this study, 12.5 % koji from *Aspergillus oryzae* W 215 plus *Halobacterium salinarium*, added to fish and 25 % salt mixture, then left to ferment for 42 days, showed good odor and flavor fish sauce. The result also indicated that *Halobacterium salinarium* was responsible for good color, flavor, and odor fish sauce. It should be noted that these studies did not compare the experimented fish sauce with commercial fish sauce.

The production of fish sauce by means of acid hydrolysis has been studied by Department of Science (1961) and Poosaran (1986a). The hydrolysate was obtained within 7 days with high degree of protein hydrolysis. However, the color and odor of fish sauce is not good when compared to commercial fish sauce.

Boonpan (2002) produced crude and purified ribonuclease from a halo-tolerant *Pseudomonas* sp. No. 3241 isolated from Thai fish sauce. This enzyme requires an optimum salt concentration of 18.0 %, and shows ability to digest RNA into 5′-GMP (an important flavor enhancer). Application of 0.5 % (w/w) crude enzyme during fermentation enhances good flavor and odor production in fish sauce. Bovornreungroj (2005) produce halophilic protease (HP) from *Halobacterium salinarum* PB407 isolated from fish sauce, then apply to fish sauce fermentation. Ribonuclease from a halo-tolerant *Pseudomonas* sp. No. 3241 (HR) was also applied during fermentation. The conclusion was that fish sauce with added 1.5 % HP or 0.5 % HR at the first period of fermentation showed a potential to compete with conventional fish sauce as it took only 6 months to ferment and showed good sensory scores.

The reviewed studies showed the potential of the addition of selected enzyme as well as koji to accelerate fish protein digestion in the first period of fish sauce fermentation, then halo-tolerant bacteria promote good odor and flavor formation in the last period; this will reduce the fermentation time compared to the conventional fermentation. However, all of the experiments were carried on laboratory scale. No report has been published as for a commercial trial.

11.5   Nutritional Values of Thai Fish Sauce

Thai fish sauce Nutrition Facts per serving size of 15 mL obtained from two different commercial brands are shown in Table 11.1. The sauce contained of 2 g protein, no fat and low carbohydrate (indicated as sugar), high sodium content (1620–1190 mg), and small amount of calcium and iron. Small amounts of vitamins, i.e., niacin, vitamin B_6, vitamin B_{12}, and folic acid are declared in some fish sauce nutrition fact tables. Vitamin B_{12} is actually produced by microorganisms during fish fermentation. Although fish sauce contains high protein, it is not promoted as a good protein source because it contains a high salt content which is not considered to be desirable in healthy diets.

Table 11.1 Nutrition Facts for Thai fish sauce (15 mL amount per serving)

Nutrition facts		Nutrition facts	
Brand A fish sauce		Brand B fish sauce	
Serving size 15 mL		Serving size 15 mL	
Amount per serving		Amount per serving	
Calories 10	Calories from fat 0	**Calories** 10	Calories from fat 0
% Daily value[a]		**% Daily value**[a]	
Total fat 0 g	**0 %**	**Total fat** 0 g	**0 %**
Saturated fat 0 g	**0 %**	Saturated fat 0 g	**0 %**
Cholesterol 0 mg	**0 %**	**Cholesterol** 0 mg	**0 %**
Sodium 1620 mg	**68 %**	**Sodium** 1190 mg	**50 %**
Carbohydrate as sugar 1 g	**1 %**	**Carbohydrate as sugar** 1 g	**1 %**
Dietary fiber 0 g	**0 %**	Dietary fiber 0 g	**0 %**
Protein 2 g	**4 %**	**Protein** 2 g	**4 %**
Vitamin A 0 %	• Vitamin C 0 %	Vitamin A 0 %	• Vitamin C 0 %
Calcium 0 %	• Iron 2 %	Calcium 1 %	• Iron 2 %

[a]Percent daily values are based on a 2000 cal diet. Your daily values may be higher or lower depending on your calorie needs

References

Areekul S, Thearewibul R, Matrakul D (1974) Vitamin B_{12} contents in fermented fish, fish sauce and soya-bean sauce. Southeast Asian J Trop Med Public Health 5:461

Boonpan A (2002) Studied on production and characterization of halophilic ribonuclease from halotolerant *Pseudomonas* sp. For fish sauce fermentation. Master of Science Thesis, Kasetsart University, Thailand (in Thai)

Bovornreungroj P (2005) Selection of extremely halophilic bacteria producing halophilic proteolytic enzyme for fish sauce production. Ph.D Dissertation, Kasetsart University, Thailand (in Thai)

Chaiyanan S (2000) Selection of bacterial strains for improvement of fish sauce fermentation. Ph.D Dissertation, University of Maryland

Chaiyanan S, Chaiyanan S (1983) Role of bacteria in flavor and odor development of Thai fish sauce. J Ind Edu Sci 2(2):1–13 (in Thai)

Chaiyanan S, Chaychotcharoen N (1987) Proteolytic activities of 10% salt tolerant bacteria. J Sci Ind Edu KMITT Thonburi 6:115–125 (in Thai)

Chaiyanan S, Chaiyanan S, Maugel T, Huq A, Robb FT, Colwell RR (1999) Polyphasic taxonomy of a novel *Halobacillus, Halobacillus thailandensis* sp. Nov. isolated from fish sauce. Syst Appl Microbiol 22:360–365

Codex Standard 302-2011. http://www.codexalimentarius.net/download/standards/11796/CXS_302e. pdf

Department of Science (1961) Production of fish sauce by rapid method. Res Rep Dep Sci 24:55–56 (in Thai)

FDA (2000) Food and Drug Administration Compilation of Laws, Rules and Regulations, Ministry of Health, Administrative Order and Guideline to Fish Sauces, 203 p

Garby L, Areekul S (1974) Iron supplementation in Thai fish sauce. Ann Trop Parasitol 68:467–476

Gildberg A, Shi XQ (1994) Recovery of tryptic enzymes from fish sauce. Process Biochem 29:151–155

Gongtip P (1965) The study of fish sauce fermentation using enzyme bromelain. B.Sc. Research, Kasetsart University, Thailand (in Thai)

Hiraga K, Nishikata Y, Namwong S, Tanasupawat S, Takada K, Oda K (2005) Purification and characterization of serine proteinase from a halophilic bacterium, *Filobacillus* sp. RF2-5. Biosci Biotechnol Biochem 38:38–44

Jongsereejit B (1990) Production of koji for fish sauce production. Master of Science Thesis, Kasetsart University, Thailand (in Thai)

Klomklang W (1995) Study on fermentation of fish sauce by using halophilic bacteria with koji. Master of Science Thesis, Kasetsart University, Thailand (in Thai)

Liptasiri S (1975) The study of some characteristics of bacteria isolated from Thai fish sauce. Master of Science Thesis, Kasetsart University, Thailand (in Thai)

Lopetcharat K (1999) Fish sauce: the alternative solution of pacific whiting and its by-products. Master of Science Thesis, Corvallis, Oregon State University

Lopetcharat K, Park JW (2002) Characteristics of fish sauce made from Pacific whiting and surimi by-products during fermentation stage. J Food Sci 67:511–516

Ministry of Public Health (2000) Notification of the ministry of public health No. 203/2543 Re: Fish Sauce www.usdathailand.org/uselectfl/rprt/18.doc

Mittranond C (1983) Experimental Fish sauce fermentation using enzymes and halophilic bacterial cultures. Master of Science Thesis, Mahidol University, Thailand (in Thai)

Muangthai P, Nakthong P (2014) Detection of some biogenic amines content in Thai sauces. IOSR J Appl Chem 7(7):53–59

Phithakpol B, Varanyanond W, Reungmaneepaitoon S, Wood H (1995) The traditional fermented foods of Thailand. Institute of Food Research and Product Development, Kasetsart University, Thailand and ASEAN Food Handling Bureau, Malaysia

Poosaran N (1986a) Fish sauce I: acid hydrolysis at ambient temperature. Songklanakarin J Sci Technol 8:43–46

Poosaran N (1986b) Fish sauce II: enzyme hydrolysis. Songklanakarin J Sci Technol 8:205–208

Raksakulthai N, Haard NE (1992) Correlation between the concentration of peptides and the flavor of fish sauce. ASEAN J Food Sci Technol 7:86–90

Raksakulthai N, Lee YZ, Haard NF (1986) Effect of enzyme supplement on the production of fish sauce from male capelin *Mallotus villosus*. Can Inst J Food Sci Technol 19(1):28–33

Saengjindawong M, Winitnuntarat S (1984) The Study of Type and number of bacteria during the first period of fish sauce fermentation. J Fishery 33(1):69–72 (in Thai)

Saisithi P (1967) Studies on the origins and development of the typical flavor and aroma of Thai fish sauce. Ph.D. Dissertation, University of Washington

Saisithi P (1968) Odor and flavor of Thai fish sauce. J Fishery 21(3):467–474

Saisithi P (1987) Traditional fermented fish products with special reference to Thai product. ASEAN J Food Sci Tech 3:3–10

Saisithi P (1994) Traditional fermented fish: fish sauce production. In: Martin AM (ed) Fisheries processing: biotechnological application. Chapman and Hall, London, pp 111–131

Saisithi P, Kasemsarn B, Liston J, Dollar AM (1966) Microbiology and chemistry of fermented fish. J Food Sci 31:105–110

Samittasiri K (1986) Halophilic bacteria in fish sauce fermentation. Master of Science Thesis, Kasetsart University, Thailand (in Thai)

Sirighan P, Raksakulthai N, Yongsawatdigul J (2006) Autolytic activity and biochemical characteristics of endogenous proteinases in Indian anchovy (*Stolephorus indicus*). Food Chem 98(4): 678–684

Suntinanalert P (1979) Role of microorganisms in the fermentation of nam pla in Thailand: relationship of the bacteria isolated from nam pla produced from different geographical localities in Thailand. Master of Science Thesis, Mahidol University, Thailand (in Thai)

Suwanik R (1977) Iron and iodine fortification of common salt and fish sauce. Research report, Medical Association of Thailand (in Thai)

Suwunnasart P (1966) Comparison of the effect of commercial enzyme and crude papain in fish sauce fermentation. Department of Fishery, Kasetsart University, Thailand (in Thai)

Thongthai C, Siriwongpairat M (1978) Changes in the viable bacterial population, pH and chloride concentration during the first month of nam pla (fish sauce) fermentation. J Sci Soc Thailand 4:73–78

Thongthai C, Mcgenity TJ, Suntinanalert P, Grant WD (1992) Isolation and characterization of an extremely halophilic archaeobacterium from traditionally fermented Thai fish sauce (nam pla). Lett Appl Microbiol 14:111–114

Chapter 12
Yunnan Fermented Bean Curds: Furu (Lufu)

Qi Lin*, Sarote Sirisansaneeyakul*, Qiuping Wang, and Anna McElhatton

Contents

*Author contributed equally with all other contributors.

Q. Lin (✉)
Faculty of Food Science and Technology, Yunnan Agricultural University,
Kunming 650201, P.R. China
e-mail: lqyncn@aliyun.com

S. Sirisansaneeyakul
Department of Biotechnology, Faculty of Agro-Industry, Kasetsart University,
50 Ngam Wong Wan Road, Chatuchak, Bangkok 10900, Thailand
e-mail: sarote.s@ku.ac.th

Q. Wang
Faculty of Food Science and Technology, Yunnan Agricultural University,
Kunming 650201, P.R. China

Department of Biotechnology, Faculty of Agro-Industry, Kasetsart University,
Bangkok 10900, Thailand
e-mail: soffywang87@gmail.com

A. McElhatton
Faculty of Health Sciences, University of Malta, Msida, Malta
e-mail: anna.mcelhatton@um.edu.mt

© Springer Science+Business Media New York 2016
A. McElhatton, M.M. El Idrissi (eds.), *Modernization of Traditional Food
Processes and Products*, Integrating Food Science and Engineering Knowledge
Into the Food Chain 11, DOI 10.1007/978-1-4899-7671-0_12

12.1   Introduction

Historical records regarding early agriculture in China report that the Emperor Xuanyuan Huangdi observed the climatic variation of the seasons and cultivated five kinds of crops: panicgrass (*Panicum antidotale*), broomcorn millet (*P. miliaceum*), beans, wheat (*Triticum aestivum*), and rice (*Oryza sativa*). This early agricultural activity was first described some 4500 years ago.

Scholars generally agree that the cultivation of soybean (*Glycine max* (L.) Merrill) first started in China with first domestication traced to the eastern half of North China in the eleventh century B.C. or there abouts. Soybean has been one of the five main plant foods of China along with rice, soybeans, wheat, barley, and millet. The word "Shu," as soybean is called in Chinese, can be found in many ancient Chinese books. The character meaning soybean actually was found in inscriptions on unearthed bones and tortoise shells of the Yin and Shang Dynasties some 3700 years ago.

The soybean forms an integral part of the Chinese people's diet and traditional soybean products such as bean curd (Tofu), soybean milk, dried rolls of bean milk cream, soy sauce, and so on are very popular ingredients favored by the Chinese people (Singh 2010). Fermented bean curd called "furu" is a traditional Chinese fermented product made of high quality soybean, and is highly valued due to its smooth texture, high nutritive value, sweet and fragrant taste, and reasonable price.

Furu dates back to ancient times in China. In the term "furu," the word "fu" means both fermenting or brewing, and bean curd (Tofu) which is the raw material for making furu. The word "ru" means soft or tender (Zhang 2002a, b). Therefore, "furu" represents a fermented soybean curd of soft appearance as a finished product.

As early as the fifth century, in Wei Dynasty descriptions of furu production have been found. The records include a description of how the bean curd was cut into small pieces, salted for 3–4 days, then dried for 2 days, steamed and then dried further for another day. Finally, bean curd was placed in a pot together with some liquor, fennel, and other spices, sealed and stored. In the JiaJing period (1507 C.E.– 1567 C.E.) of the Ming Dynasty, Shaoxing Furu of Zhejiang province was well known and sold to many Asian countries such as Singapore, India, Myanmar, and further afield.

Furu is a cheese-like product that contains proteins and fats that are hydrolyzed by microorganisms during fermentation and turned into peptides, amino acids, glycerols, and fatty acids that are easier to digest. This product is known as Chinese cheese (Han et al. 2001a, b) or oriental cheese because its nutritional profile is similar to that of cheese. Traditions and life habits vary in China, and similarly the types and tastes of furu differ in various parts of China and have their own descriptions and names such as Shaoxing furu in Zhejiang, Guilin furu, Shilin furu (or "lufu") in Yunnan, Tangchang furu in Sichuan to name a few.

It is said that the production of furu in Shilin has a long history dating back to Qin and Han Dynasty (221 B.C.–220 C.E.). In Tang Dynasty (618 C.E.–907 C.E.), furu products made in Shilin were often given as tribute to royal court. Such quality products have a yellow-red color, spicy taste, fragrant smell, and smooth texture. Furu is usually processed in winter, stored in spring, and sold in summer. In Yunnan people commonly eat this product with rice and often use it as a seasoning when cooking with other foods such as meat.

12.1.1 Classification of Furu

There are various types of furu on the market that are produced by various processing methods. In addition, in the fermentation process of bean curds, types of furu are also classified by their color, taste, odor, and appearance (Table 12.1).

12.1.1.1 Cured Type

Bean curd does not mold, or undergo pre-fermentation by microorganisms as compared to other types, but goes directly into the post-fermentation stage. Additional components such as *mian gao qu* (starter cultured flour prepared with *Aspergillus oryzae*), *hong qu mi* (red rice, starter culture and pigments made from rice inoculated with *Monascus* van Tieghem) and rice white liquor or yellow liquor are added to promote the biochemical reactions of curing type furu. Such a process is relatively simple; it is not labor intensive and does not require complicated equipment. However, because of low levels of proteinase, the fermentation time needed to produce this type of furu is long with the final product having a rough texture and containing low levels and limited profiles of particular amino acids.

Table 12.1 Classification of furu with tastes and types

Furu types	Descriptive characteristics
Red	Bean curds are treated with *Hong Qu Mi* before putting into jar for post-fermentation, end product appears with dark red color on surface and yellow inside. It tastes sweet, fresh, and alcoholic aroma as well
White	The color of products is creamy yellow, light yellow, or pale. It tastes rich ester fragrance, fresh, and smooth texture
Various tastes	Paste bean sauce is as the main supplement material added in the post-fermentation stage
Sauce	Paste bean sauce is used as the main supplement material in the post-fermentation stage
Green	It is also called "strong-smelling furu" with cyan or green bean color. Smells odorous but eats fragrant. It is fresh with a small cake shape and exquisite character

12.1.1.2   Natural Inoculation Type

Bean curd is naturally fermented by those microorganisms that are naturally found in the local environment. It is usually incubated at temperature less than 15 °C for 7–15 days until a grey-white mycelium grows on the surface of fermented bean curds, during which period various enzymes are secreted extracellularly. These enzymes give furu a smooth texture and render it rich in amino acids. The process does not require specialized equipment, but has a relatively long production period; consequently mass production throughout the year is difficult to achieve as optimal process conditions are seasonally limited by climate conditions.

12.1.1.3   *Mucor* Type

During pre-fermentation, soybean curds are inoculated with *Mucor* culture in suspension or in powder form. After 48–72 h of incubation time, the surfaces of small cakes are covered with a mass of white mycelium. These mycelial cultures not only infiltrate the bean curds, but also secrete considerable enzymes to degrade soy proteins to produce the characteristic quality and nutrition profile.

12.1.1.4   *Rhizopus* Type

The production of this type of furu involves the use of thermo-tolerant *Rhizopus* culture to ferment the bean curd. The *Rhizopus* starter culture functions like the *Mucor* culture, but at a temperature of 37 °C.

12.1.1.5   Bacterial Type

Furu is also produced through bacterial fermentation. The finished product would have undergone a high degree of protein hydrolysis that results in a smooth texture, but with poorer shape retention characteristics.

12.1.2   Nutritive Value of Furu

Furu is a fermented soybean product. It is well known that soybean is rich in protein (36–40 % protein), and fat (8 % fat; 7 % saturated, 41 % unsaturated). Soybean contains various vitamins such as thiamine, niacin, and vitamin A and numerous minerals such as calcium, phosphorus, and iron, all of which are food components needed for a healthy diet. As a kind of fermented soybean product, furu not only possesses the nutritive values of soybean with high quality protein, abundant unsaturated fatty acids such as linoleic acid, oleic acid without cholesterol, but also it is

Table 12.2 Composition of furu and yellow liquor

General nutrients of furu (in 100 g furu, dry basis)

Composition	Content	Composition	Content
Moisture (g)	56.3	Calcium (mg)	231.6
Protein (g)	15.6	Phosphorous (mg)	301.0
Fat (g)	10.1	Iron (mg)	7.5
Sugar (g)	7.1	Zinc (mg)	6.89
Crude fiber (g)	0.1	Vitamin B_1 (mg)	0.04
Ash (g)	1.12	Vitamin B_2 (mg)	0.13
Cholesterol (g)	ND	Niacin (mg)	0.5

Amino acids content in furu (g/100 g protein, dry basis) and yellow liquor (mg/kg)

Amino acids	Furu	Yellow liquor	Amino acids	Furu	Yellow liquor
Alanine	10.0	340	Serine	2.3	200
Glutamic acid	0.6	420	Valine	0.3	320
Leucine	8.8	310	Asparagine	5.1	ND
Proline	2.4	400	Histidine	1.4	80
Tyrosine	2.2	230	Methionine	0.7	40
Arginine	2.1	390	Threonine	2.0	130
Glycine	4.4	290	Cystine	0.4	120
Lysine	7.0	180	Isoleucine	4.8	210
Phenylalanine	4.6	230	Tryptophan	0.6	10
Aspartic acid	ND	290			

Based on Zhang and Liu (2002a, b), and Dong and Xu (2003)

characterized by high content of calcium and easy ingestion and digestion. In the furu making, process, water is used as solvent to extract soybean protein. Brine solution or a salty liquor and gypsum which contains abundant calcium are used as coagulating agents; furu therefore contains added calcium. In addition, the amino acids content of product increases due to fermentation catalyzed by enzymes secreted from microorganisms to hydrolyze soybean protein to amino acids and peptides. The typical compositions of furu are shown in Table 12.2.

12.2   Fermentation Processes

12.2.1   Raw Materials and Supplements of Furu Production

The main raw material of furu production is soybean. Defatted soybean is traditionally also used in some regions. The other ingredients needed include salt, yellow distilled liquor, Hong qu (red rice, starter culture and pigments made from rice inoculated with *Monascus* van Tieghem), Mian qu (starter cultured flour prepared with *Aspergillus oryzae*), sugar, and various spicy substances.

12.2.1.1   Main Raw Materials

There are three main raw materials for making furu as follows.

Soybean

Soybean is the preferred raw material for the production of furu. When processed this resultant furu has a particular soft, glutinous, fine texture, and good taste.

China is one of the main soybean production countries with at least 10,000 varieties characterized and known to have different attributes and requisites such as planting season, growth and development, color of seed capsules, degree of evolution, shape and size of seeds, and different uses and application of the soybeans. For quality attributes furu should be made from soybeans that are high in protein. In China the main soybean varieties are known to have a typical composition of over 36 % protein and 18 % fat. Chemical compositions of four main varieties of soybeans are summarized as follows: (1) Yellow bean, 9.4–10.2 % moisture, 14.22–19.88 % crude fats, 36.06–41.96 % crude protein, 18.70–30.09 % carbohydrate, 3.10–6.81 % crude fiber, and 3.77–5.67 % ash, (2) Cyan bean, 9.16–12.64 % moisture, 15.98–18.30 % crude fats, 35.58–39.81 % crude protein, 19.31–25.68 % carbohydrate, 4.98–11.67 % crude fiber, and 4.28–4.89 % ash, (3) Black bean, 13.96 % moisture, 19.85 % crude fats, 36.59 % crude protein, 21.33 % carbohydrate, 4.05 % crude fiber, and 4.23 % ash, and (4) Brown bean, 11.36–13.80 % moisture, 19.10 % crude fats, 33.44–37.55 % crude protein, 20.55 % carbohydrate, 4.16 % crude fiber, and 4.12 % ash (Dong and Xu 2003). The yellow and cyan beans are often selected because the soybean to furu yield ratio is considered to be good and is, therefore, the preferred bean for the production of furu. Black or brown beans on the other hand produces poor quality black furu that has a hard texture and lower production yield.

Defatted Soybean

Defatted soybean is the by-product obtained after the removal of the soybean's oil. Due to different methods of oil extraction, there are two types of soybean residues produced, namely bean cake and bean draff.

Bean Cakes

Oil extraction using hot or cold conditions can be used. Heat extraction carried out at relatively high temperature produces a higher oil yield, but is associated with thermal damage to soybean proteins. While cold extraction at ambient temperature yields less oil and causes less damage to soybean proteins. Different extraction methods not only result in different quality of bean cakes as raw materials for the production of furu, but also lead to a different chemical composition of resulted bean cakes. Cool squeezing cake generally has 12.00 % moisture, 46.45 % crude protein, 6.12 % crude fat, 20.64 % carbohydrate, and 5.49 % ash. While heat squeezing cake typically has 11.00 % moisture, 47.94 % crude protein, 3.14 % crude fat, 22.84 % carbohydrate, and 6.31 % ash (Dong and Xu 2003).

Bean Draff

Soybeans are primarily heated to regulate the water content and then pressed to a flat shape. Organic solvent extraction is applied to remove oil from the aforementioned pretreated soybeans. Defatted soybean, known as bean draff, is a by-product of oil extraction. The bean draff is exposed to vacuum and low temperature to remove organic solvent. In China this draff is considered to be a good quality raw material for furu production. It contains less fat and moisture, and therefore has higher protein content. The composition of defatted soybean after solvent extraction is shown as follows: 7–10 % moisture, 46–51 % crude protein, 0.5–1.5 % crude fat, 19–22 % carbohydrate, and 5.00 % ash (Dong and Xu 2003).

12.2.1.2   Supplement Materials

Water

Water used in furu production has to meet drinking water quality standards. Table 12.3 summarizes the effects of water hardness on tofu quality and yield.

Coagulating Agents

The main component of furu is soybean protein. Coagulatants are substances which cause aggregation of the soy protein. There are two types of coagulating agents used in Tofu making which are salts and organic acids. The former gives a higher production yield than the latter, while organic acids as coagulatants make tofu taste remarkably smooth. Two types of coagulating agents can be used together to compensate from the other strong points to offset one's weakness.

Table 12.3 Effects of water hardness on tofu quality and yield

Types of water	Protein in bean milk (%)	Yield of tofu (%)
Soft water	3.71	45.0
Pure water		47.5
Well water	3.40	30.0
300 mg calcium/L water	2.40	26.5
300 mg magnesium/L water	2.00	21.5

Based on Dong and Xu (2003)

Brine Solution

Brine solution is bitter water solution of bromide, magnesium, and calcium salts that remain after sodium chloride is crystallized out of seawater. Usually the concentration of brine solution is 25–27 °Bé (as soluble solids). It should be diluted to 16 °Bé when added to bean milk as a coagulatant. The main components of the brine solution include $MgCl_2$ (~30–40 %), $NaCl$ (<2 %), $MgSO_4$ (<3 %), and KBr. It is widely used in Tofu production because Tofu coagulated by a brine solution possesses good fragrance and taste.

Gypsum

Gypsum is a widespread white or yellowish mineral. It is sparingly soluble in water, so it reacts with protein slowly. Tofu made by gypsum is smooth, fine, and retains water well with moisture levels in Tofu being about 88–90 %. There are different types of gypsums, with the hydrated type ($CaSO_4 \cdot 2H_2O$) considered to be the best for Tofu making because of its tendency to cause a quicker coagulation of the soya milk. Plaster of Paris ($CaSO_4 \cdot 0.5H_2O$) has a slower action, while its dehydrated type ($CaSO_2$) has almost no function at all.

Gypsum should be heated then mashed to break it down into a powder and finally added to water to make solution. Tofu made with gypsum does not have the specific characteristics of that made using a brine solution.

Gluconolactone

Gluconolactone is easily mixed with bean milk. At 66 °C, gluconolactone converts to gluconic acid which coagulates bean milk while retaining/binding water, the resultant Tofu is smooth and tender. It should be noted that the yield obtained by this process is high. At pH and high temperature, the coagulation is rapid, when the temperature is 100 °C and pH between 6 and 7, the recovery yield of Tofu reach 80 % and 100 %, respectively. In practice, dosage of gluconolactone is usually 0.01 mol/L, which results in a Tofu that is described as having a good flavor.

Gluconolactone tastes sweet and dissolves in water easily. However after conversion to gluconic acid, it becomes sour which also may make the Tofu taste sour. Mixed coagulatants may therefore be considered as follows: gluconolactone 20–30 % mixed with gypsum 70–80 %, or a mixture of gluconolactone and magnesium chloride. These mixtures will improve the flavors of Tofu, but have to be applied quickly due to their tendency to quickly cause coagulation.

Besides the brine solution, gypsum, and gluconolactone mentioned above, calcium acetate, calcium lactate, and calcium gluconate are also alternatively used as coagulating agents.

Salt

Salt is one of the main and indispensable supplement materials in furu production, which has a crucial function during all the process of fermentation. It has various functions, it gives a desirable salty taste to the product, it combines with amino acids to develop the taste, it inhibits some microorganisms and prevents deterioration of LUFU. The salt used in furu production has to be dry and of good quality and free of impurities.

Seasonings

Seasonings are essentially used to improve flavors and increase the number of various furu products. Distilled spirits or liquors and some spices are particularly used as seasonings in furu production.

Alcoholic Seasonings

Yellow liquor: This is a special local Chinese product made using a unique brewing method and flavoring. It is one of the oldest known liquors not only in China but also in the world. It is mostly produced in Jiangsu, Zhejiang, Shanghai, Jiangxi, and Taiwan. Yellow liquor, namely *Huang Jiu*, is made from glutinous rice by fermentation involving Rhizopus and other yeasts. It is characterized by a low alcoholic content (16 %vol.), mellow taste, and strong aroma. Furu's fragrance, specific flavor, and grade are enhanced by adding proper amounts of yellow liquor to furu during fermentation of the bean curd. A typical yellow liquor (*Huang Jiu*) has 16 % alcohol, 7 % soluble substances, and 5 kJ/L calorie (Dong and Xu 2003). Amino acids content of yellow liquor is shown in Table 12.2.

   Distilled spirit: This product is made from cereals by yeast fermentation and distillation. It has a 50–60 % alcohol content and is colorless. It can be drunk directly or diluted to use in the production of furu as a replacement of yellow liquor or added to *laochao*, which is an alcoholic rice paste.

Spices

Spices are usually added in the post-fermentation stage of furu production. Kinds and quantities added vary depending on type of furu, and may also include Chinese pepper, fennel, cinnamon, dry orange peel, ginger, hot pepper, and other traditional ingredients. These spices have components which develop different aromas and flavors when mixed into the furu. Some of components in spices have antimicrobial activity and contribute to the preservation of the product.

Starter Cultures

Hong Qu (Red Rice)

Red rice is made from rice fermented by *Monascus* and is an important supplement material used in the post-fermentation stage. It provides the product with pigmentation, flavors, and fragrances. It enhances the amino acids and peptides profiles and is described as having antiseptic properties. It is reported to promote appetite due to the high content of amylase, which can efficiently digest starch. The red pigments consist of monascorubrin ($CH_{22}H_{24}O_5$) and monascoflavin ($C_{17}H_{22}O_4$) which are sparingly soluble in water but easily soluble in organic solvents. It should be noted that the red color will be fade when exposed to sunlight.

Mian Qu

It is semi-finished soy sauce made from flour fermented by *Aspergillus oryzae*. Because *Aspergillus oryzae* and other organisms secrete plenty of proteinase and amylase, addition of Mian qu to the post-fermentation stage will not only improve flavors and fragrances, but will also speed up the ripening of furu.

12.2.2 Techniques of Furu Production

The traditional production of furu is through bean processing and fermentation utilizing soybeans as the starting material. Classification of furu production with processing methods is summarized in Fig. 12.1. The finished product is a unique oriental fermented food that is a good source of amino acids. The processing of bean curd from soybeans and fermentation of bean curd for furu production are summarized in Fig. 12.2.

12.2.2.1 Processing Stages of Tofu Production

Soaking

Soybean proteins are located within bean tissue as colloids. Soybean has a hard shell and dense tissue which prevents release of proteins. Soaking therefore causes beans to absorb water, to increase the hydration of protein, loosen the tissue structure, which then causes cell walls to swell and break down. The softened shells then easily disintegrate causing proteins to dissolve in the water to form a "milk." When soaking, attention must be given to:

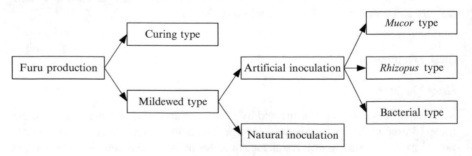

Fig. 12.1 Classification of furu production with processing methods

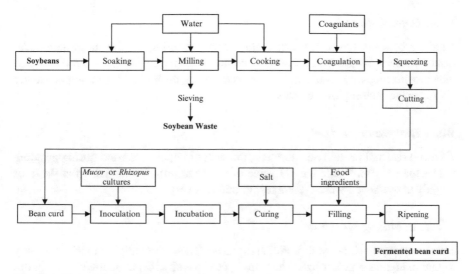

Fig. 12.2 Flow diagram of furu production from soybeans

Water Quantity Added

The volume added has to be controlled in relation to the composition of the bean. The appropriate proportion between water and bean is (3–4):1 for soaking.

Quality of Water

Water has to meet drinking water quality standards.

Soaking Time

Soaking time is closely associated with water temperature. While water temperature is 5–10, 15–20, and 25–30 °C, appropriate soaking time is 18–22, 10–15, and 5–8 h, respectively (Dong and Xu 2003). Excessive soaking time causes protein loss by enzymatic hydrolysis and growth of microorganisms which lower the pH of water and influence proteins recovery. While insufficient soaking results in underhydrated beans which also influences the recovery of proteins.

Milling

Soaked beans are mechanically milled to break down the cell tissues to promote the formation of the bean milk. Factors affecting the protein yield include the extent of grinding, water quantity, and water temperature and pH value.

Grinding

Diameter of the soybean globin is about 5–15 μm, grinding should be sufficient to promote dissolution of soybean proteins in water.

Water Quantity Added

Addition of water lowers the temperature during grinding to prevent protein denaturation caused by the heat produced, and promotes the hydration of protein which aids protein recovery. Water quantity added during milling and grinding is usually six times the quantity of the bean.

Water Temperature and pH

To avoid thermal protein denaturation, the water temperature used during grinding is less than 10 °C. While the pH value of the bean milk is maintained at seven or slightly above seven which promotes protein recovery.

Choice of Milling Machines

Stone milling machine were traditionally used. These are simple in structure, easy to maintain, and small in size. These mills produce significant heat which degrades protein thus decreasing the production yield of soybean curd. Currently, grinding wheels that are used by bean production factories generate much less heat than the stone milling machines.

Soybean Milk Sieving

Soy bean milk sieving separates residues from water soluble substances which are a good basis to manufacture high quality Tofu. The separation equipment used generally consists of a sieving machine and centrifuge. Water should be added three to four times during grinding. 100 kg bean so processed should yield 1100 kg bean milk at a concentration of 5.5–6.0 °Bé.

Cooking of Bean Milk

Bean milk cooking causes denaturation of protein and promotes further coagulation of protein. The Tofu formed is white, elastic, tender, and shiny and has good water retention. When the milk is cooked some denaturation will occur, otherwise even when coagulants are added the protein will not curdle.

Cooking eliminates factors harmful to body such as agglutinins, trypsin inhibitor and saponin inhibitor, and sterilizes the bean milk. The favored condition of this thermal treatment is 100 °C for 5 min.

Dian Jiang (Addition of Coagulants)

Addition of coagulants to bean milk to cause curdling is called *dian jiang*. It is one of the key steps used to manufacture high quality bean curd. Factors during manipulation such as density of soybean milk and brine solution, temperature and milk pH value are monitored.

The density of soybean milk should be strictly controlled to a level of 4–5 °Bé. Yield and quality of Tofu will be greatly affected if the milk is too low or too high density.

The density of brine solution directly affects the yield and quality of Tofu. The appropriate concentration is 14–16 °Bé.

Temperature affects the curdling effects of protein. High temperature causes faster curdling which results in a loose/open structure of bean curd, while low temperature causes slower curdling which results in incomplete curdling and significant loss of protein. The temperature is usually controlled within 85–90 °C.

Bean milk pH value also affects the curdling of protein. The appropriate value is pH 6.7–6.8.

When brine solution is used as coagulatant, it should be added to hot milk slowly while stirring the milk. The stirring speed should be lower when it is on the point of coagulation. Stirring should be maintained until complete coagulation is reached. The whole *dian jiang* process usually takes 5 min. To form perfect curd, another 15–20 min is needed for bean curd to complete the protein and coagulatant interaction.

Squeezing

After the *dian jiang* process, the curd sinks and squeezing is performed. This process firms the curds and removes excessive yellow water from curd. To produce quality dispersed protein three factors during squeezing are monitored, namely pressure, temperature, and duration. If temperature of curd is very low, even if the applied pressure is high enough, the combination between protein gelatins is still not compact, and the curd flans formed tend to be fragile and will break. Squeezing needs some time; if squeezing time is not sufficient, the typical shape will not form. If on the other hand the squeezing time is too long, too much water inside curd will be squeezed out. Squeezing pressure is also important. Without sufficient pressure, water cannot be squeezed out from the bean curd and the matrix produced is loose. While if pressure is too high the gelatinous protein tissues will break down. The squeezing appropriate temperature, pressure, and duration are therefore applied as follows: temperature above 65 °C, 15–20 kPa pressure, and a duration of 15–20 min, respectively. Water content in curd after squeezing is 66–71 %, protein content should be above 14 %.

Cutting

After squeezing, the curd is cut into smaller pieces according to the product standard. This is the last step of manufacturing bean curd flans.

12.2.2.2   Furu Fermentation

Furu fermentation includes two stages namely pre-fermentation and post-fermentation. The former refers to inoculating and incubating *Mucor* or *Rhizopus* on bean curd flans and promoting the growth of organisms sufficiently which form a layer of tough and slender mycelium on the surface of soybean curds. Because of the mycelial growth, large amount of enzymes are produced such as proteinases, amylases, and lipases to hydrolyze proteins and other substances in post-fermentation. In pre-fermentation some proteins are already hydrolyzed. Post-fermentation is an anaerobic process as is the maturation process. In this period, due to the contribution of molds, yeasts, bacteria, and supplement materials, many biochemical reactions would have taken place such as protein hydration, starch decomposition to sugars, organic acids fermentation and formation of esters. These biochemical reactions produce desirable substances which form the specific color, texture, flavors, and fragrances of furu.

Pre-fermentation

Pre-fermentation is the process of growth of mycelium. Natural fermentation and artificial inoculation are both possibly used in the production of furu. Pre-fermentation with mold inoculation is introduced as shown in Fig. 12.3.

Inoculation

Two types of starter cultures are used in production. The starters are either applied as a suspension or in solid form. When a spore suspension is used, the starter culture is diluted and spread onto the surface of bean curd. When the solid starter culture is used, it is mixed with carriers such as rice powder or corn power, then scattered uniformly onto the surface of bean curds.

Fig. 12.3 Flow diagram of pre-fermentation for furu production

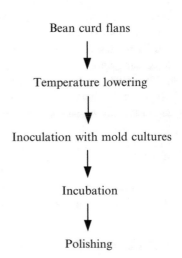

Bean curd flans

↓

Temperature lowering

↓

Inoculation with mold cultures

↓

Incubation

↓

Polishing

Incubation

Every piece of bean curd cakes is put standing against cage drawers. Spaces between cakes should be maintained to ensure good ventilation and temperature regulation. Cage drawers on which bean curds stand are piled up one by one and upper ones are covered. Appropriate room temperature for incubation ranges between 20 and 25 °C, and should not exceed 28 °C. During incubation, to regulate temperature between cage drawers and keep organism growth uniform, drawers should be moved up and down two to three times. The first moving is usually carried out 22 h after incubation when the growth of mycelium has started and goes on for the next 6 h to about 28 h, at this point the mycelium promotes ripeness and a second moving of drawers of curs starts. The third moving depends on the temperature and mycelial growth situation. After 44–68 h (depending on temperature) incubation, when the bean curds are covered with white or light yellow-white mycelium, the pre-fermentation finishes and goes quickly into the following steps.

Mycelium Rubbering or Polishing

This process refers to the manual flattening of the blooming active mycelium and the separation of curd cakes from each other. It is an important operation required to keep the shape of the cakes.

Post-fermentation

After pre-fermentation, *mao-tofu* which is the pre-fermented soybean curds covered with a sheet of white mycelium, is ready for further fermentation. Curing with salt and addition of food ingredients are applied at this stage of production. Mixed cultures of yeasts and bacteria rather than molds play their important role for ripening the pre-fermented bean curds. Figure 12.4 shows final steps of post-fermentation for ripening furu.

Curing

Curing begins immediately after pre-fermentation finishes. With curing water from both mycelium and *mao-tofu* is separated out, which makes *mao-tofu* become tough. Salt has anti-spoilage and preservation properties. High concentration of salt will inhibit the enzyme activity of proteinase which results in delaying various hydrolysis reactions that keep the furu's matrix integrated and therefore prevents it breaking down. Salt has organoleptic properties and acts as seasoning.

The amount of salt used ranges from 57.5 to 65 kg per 10,000 pieces of cakes (4.1 × 4.1 × 1.6 cm) depending on seasons and curing temperature. Duration of curing ranges from 6 to 13 days depending on different regions and seasons. When curing finishes the final salt content of the cakes is 11–14 %. Finished cured products should be laid for a night to let the cakes dry.

Fig. 12.4 Flow diagram of
post-fermentation for furu
production

Filling, Sealing, and Post-fermentation

Salted cakes are put into jars or bottles while supplement materials are prepared together with the cakes. Different types and varieties of furu have different combinations of supplement ingredients added. The most important ingredients are liquors and salt. *Hong qu mi* is indispensable in the production of red furu. Other seasonings and spices such as Chinese pepper, fennel, cinnamon, dry orange peel, ginger, and hot pepper are added according to the products' specifications. After jar loading finishes, the jars should be sealed tightly and goes into post-fermentation. It takes about 6 months at ambient temperature until furu ripens.

12.3   Research and Development

12.3.1   *Research on Microflora of Furu Fermentation*

Microorganisms are important in the production of furu. Without the involvement of microorganisms furu cannot be produced. Traditional natural inoculation and fermentation were used in furu production. The isolation, screening, identification, and structured research of furu fermentation microorganisms started in the 1940s when a microbiologist by the name of Fan Xinfang isolated *Mucor* from the Wutongqiao furu of Sichuan. Since then, research on furu fermenting microorganisms began. So far, microorganisms isolated from furu are summarized in Table 12.4.

   Among these microorganisms, *Mucor* is predominant and widely applied due to its good performance in the furu production. Frequently used species include *M. wutung kiao* and *Actinomucor elegans*. *Rhizopus* is also used because they withstand

Table 12.4 Potential microorganisms isolated from furu products

Microorganism	Production region
Mucor sufu	Saoxin, Zhejiang; Suzhou, Jiangsu
M. rouvanus	Jiangsu
M. wutung kiao	Wutongqiao, Sichuan
Mucor so	Taiwan, Guangdong, Guilin
M. racemosus	Taiwan, Sichuan
Actinomucor elegans	Beijing, Taiwan, Hong Kong
M. hiemalis	Taiwan
M. feavus	Wutongqiao, Sichuan
Monascus purpureus	
Rhizopus liquefaciens	Jiangsu
Aspergillus oryzae	Jiangsu; Wutongqiao, Sichuan
Penicillium sp.	Jiangsu
Alternaria sp.	Jiangsu
Cladosporium sp.	Jiangsu
Bacillus sp.	Wuhan
Micrococcus luteus	Heilongjiang
Saccharomyces	Jiangsu; Wutongqiao, Sichuan
Bacterium sp.	
Streptococcus sp.	

Based on Dong and Xu (2003)

high temperature. Han et al. (2003) compared the properties of the usual starter *Actinomucor elegans* and the potential alternative starter *Rhizopus oligosporus*. The effects of temperature and relative humidity on growth rate of *Actinomucor elegans* and *Rhizopus oligosporus* were optimized at 25 °C, 95–97 % RH, and 35 °C, 95–97 % RH, respectively. Yields of protease (108 U/g pehtze or *mao-tofu*, fresh bean curd overgrown with mold mycelia), lipase (172 U/g), and glutaminase (176 U/g) by *A. elegans* were maximized after 48 h at 25 °C and 95–97 % RH, and for α-amylase (279 U/g pehtze) and α-galactosidase (227 U/g) at 30 °C and 95–97 % RH after 48 and 60 h of incubation. Highest protease (104 U/g pehtze) and lipase (187 U/g) activities of *R. oligosporus* were observed after 48 h at 35 °C and 95–97 % RH, while maximum α-amylase (288 U/g pehtze) and glutaminase (187 U/g) activities were obtained after 36 h at 35 °C and 95–97 % RH. Maximal α-galactosidase activity (226 U/g) by *R. oligosporus* was found after 36 h at 30 °C and 95–97 % RH. It is thought that *R. oligosporus* is a potential alternative to *A. elegans* as furu pehtze starter during hot seasons.

12.3.2   Research on Mechanisms of Post-fermentation

Post-fermentation of furu is the main stage in which typical flavor forms and protein degradation takes places. In recent years, research on biochemical changes during post-fermentation of furu has been given more attention. Yu and

Chen (2001) revealed that content changes of soluble proteins and amino acids during the post-fermentation reflected directly the ripening situation of furu. When soluble proteins reach 18–20 % (dry basis) and amino acids exceed 0.5 %, furu is considered to be ripe. The research of Lu and Sun (2003) demonstrated that texture, structure, acidity, and chemical composition of furu changed as the post-fermentation progressed. Furthermore, these changes mainly happened within the last 40 days of fermentation. During this fermentation period, viscosity of the cakes decreased rapidly, acidity increased gradually, free amino acids content raised as fermentation went on, amylase activity was stable while proteinase and lipase activities varied. Ma and Han (2003) studied the effects of NaCl on textural changes and protein and lipid hydrolysis during the post-fermentation stage of furu. The results showed that elasticity and hardness of product with high content salt (14 %) increased while viscosity decreased. The amount of free amino acids quickly increased as a consequence of rapid hydrolysis of protein of product with low salt content. In contrast, hydrolysis progressed slowly for high salt content product. Subsequently, furu production with 8 % salt or below was suggested which provided a basis for low salt content production. The content and state of soybean isoflavones during the ripening stage of furu were studied by Yin et al. (2004). It was shown that the loss of isoflavones was mainly attributed to the preparation of Tofu and salting of pehtze (fresh bean curd overgrown with mold mycelia). Soybean isoflavones existed in four states during the ripening stage and their composition and amount changed during the progression of the process. The levels of aglycones increased, while the corresponding levels of glucosides decreased. The isoflavones in furu, in the form of aglycones and in the form of glucosides, accounted for 99.7 % and 0.3 % of the total, respectively. The changes in the isoflavone composition were significantly related to the activity of β-glucosidase during furu fermentation, which was affected by the NaCl content. It was also known that on both the 30th and 90th days the content of soybean isoflavone reached the maximum. All these works above provided a better understanding on the biochemical changes and mechanism during ripening stage of furu and to some extent, established a theoretic basis for the development of good practice needed to control or regulate the ripening process of furu.

12.3.3 Problems Existing and Development

In the production of furu, problematic processing and inadequate knowledge are issues that furu manufacturers of SME scale, food scientists, and technologists are currently facing. Research and development for hygienic production of furu with mixed starter cultures are needed to establish a novel clean industry of mass production.

12.3.3.1   Existing Issues

1. Production period is too long. It usually takes about 6 months to finish the ripening stage, which results in a lower productivity and affecting the better use of factory facilities, equipment, and finance (Zhang 2002a, b).
2. Most of the enterprises use the single organism *Mucor*, for example, to the furu production which does not tolerate high temperature, so that production has to be stopped during hot summer (Zhao and Zheng 1999). Furthermore, strain variation and deterioration is a reality that has to be mitigated (Li 1999).
3. Very old traditional ways of production are still being used in some enterprises resulting in inefficient manufacture that remains complex and very labor intensive (Lin 2000). Issues with quality, consistency, and hygiene of the product are difficult to maintain in such traditional environments.
4. Most furu products have high content of salt which limits their wide consumption in a health conscious society, especially for those hypertension (Zhang 2002a, b).

12.3.3.2   Future Research and Development

After years of working on furu fermentation, useful microorganisms predominant in furu production, as well as some knowledge of biochemical changes during ripening stage, have been produced. However, other issues that still need to be investigated include:

1. Single organism inoculation and fermentation and natural fermentation are characterized by their limitations (Zhang and Pu 2005). Multi-strains fermentation could complement each other with benefits to all year round production, shorter post-fermentation period and protein hydrolysis. Research on this area should be conducted.
2. Salt content directly affects the flavor and quality of furu. Products with low content of salt (5 %) deteriorate in the ripening stage of furu, while high salt content of furu hardens the product and lessens the organoleptic quality, 8 % salt is considered appropriate content for furu (Ma and Han 2003).
3. Traditional technology of furu production uses the two-stage fermentation by microbes which is lengthy and complex. The fermentation is actually an enzymatic process in which organic substances such as protein, lipid, and starch are hydrolyzed and aromas produced. Further work on the enzyme technology of furu fermenting microbes is needed (Song and Ju 2002). Purification of enzymes secreted from microbes and application to furu production will greatly enhance the production process and shorten the reduction period, as well as decrease the chance of contamination.
4. For a long time, furu has been sold in the market in the shape of a cake. This process has a lot of limitations such as rough package design which is not attractive to consumers, it is inconvenient to eat and discolors easily when the cakes

are exposed to air. The development of paste furu could be complementary to or an alternative of the traditional cake furu. Furu paste, which needs less operation by hand and has higher productivity, is characterized by uniformity and fluidity which lends itself well to various forms of packaging including packaging in bottles, soft packaging, or other forms of packaging. Products so packed would be more hygienic and competitive (Lin 2000).

References

Dong SL, Xu KS (2003) Production technology of brewing seasonings. Chemical Industry Press, Beijing

Han BZ, Beumer RR, Rombouts FM, Nout M Jr (2001a) Microbiological safety and quality of commercial sufu—a Chinese fermented soybean food. Food Control 12(8):541–547

Han BZ, Ma Y, Rombouts FM, Nout M Jr (2003) Effects of temperature and relative humidity on growth and enzyme production by *Actinomucor elegans* and *Rhizopus oligosporus* during sufu pehtze preparation. Food Chem 81(1):27–34

Han BZ, Rombouts FM, Nout M Jr (2001b) A Chinese fermented soybean food. Int J Food Microbiol 65:1–10

Li YJ (1999) Studies on *Mucor* property and the new way of furu technology seeking. J China Condiment 10:5–9, In Chinese

Lin ZH (2000) Debate on paste furu development. J China Condiment 8:27, In Chinese

Lu F, Sun SJ (2003) Studies on some components changes during furu fermentation. J China Brewing 6:14–17, In Chinese

Ma Y, Han BZ (2003) Effect of NaCl on proteins and lipids hydrolysis during furu production. J China Brewing 6:14–17, In Chinese

Singh G (2010) The soybean: botany, production and uses. CABI, Oxfordshire

Song JM, Ju HR (2002) New edition of soybean food processing technology. Shandong University Press, P.R. China

Yin LJ, Li LT, Li ZG, Tatsumi E, Saito M (2004) Changes in isoflavone contents and composition of sufu (fermented tofu) during manufacturing. Food Chem 87(4):587–592

Yu RQ, Chen WM (2001) Chemical changes during post-fermentation stage of furu production. J South China Univ Technol 5:697–67, In Chinese

Zhang Q (2002a) The present status of the production of preserved bean curd in China. J China Condiment 6:9–13, In Chinese

Zhang XM, Pu B (2005) Development and prospects of fermented soybean curd. J Food Ferment Ind 31(5):94–97, In Chinese

Zhang ZP (2002b) Study and explain of name of furu. J China Brewing 2:41–42, In Chinese

Zhao XL, Zheng XX (1999) Furu production microorganisms. J China Condiment 2:12–15, In Chinese

Chapter 13
Tempe from Traditional to Modern Practices

Hadi K. Purwadaria, Dedi Fardiaz, Leonardus Broto Sugeng Kardono, and Anna McElhatton

Contents

H.K. Purwadaria
Department of Food Technology, Swiss German University—SGU, Serpong, Indonesia

D. Fardiaz
Department of Food Science and Technology, Bogor Agricultural University—IPB,
Bogor, Indonesia

L.B.S. Kardono
Research Centre for Chemistry, Indonesian Institute of Sciences—LIPI, Serpong, Indonesia

A. McElhatton (✉)
Faculty of Health Sciences, University of Malta, Msida, Malta
e-mail: anna.mcelhatton@um.edu.mt

© Springer Science+Business Media New York 2016
A. McElhatton, M.M. El Idrissi (eds.), *Modernization of Traditional Food
Processes and Products*, Integrating Food Science and Engineering Knowledge
Into the Food Chain 11, DOI 10.1007/978-1-4899-7671-0_13

13.1   Introduction

13.1.1   Historical and Cultural Development of Tempe

Tempe or tempeh is a soybean fermented solid food product originally and traditionally made in households and cottage industries in Indonesia. Even though the word "Tempe" is thought to be originated from "tumpi" a word from old Javanese language, "Tempe" has been mentioned as it is in the old Javanese literature, Serat Centhini, circa 1814 (Astuti 1996). Serat Centhini, volume 3, documented a trip of the royal family of Sunan Giri at Central Java in the end of the seventeenth century, and cited that at one location they were served foods which were described in detail like as follows: "*Jangan menir ulur-pitik, brambang kunci sambel sinantenan, brambang jae santen Tempe, ...*" (Chicken eggs sauteed with spinach, hot sauce with coconut milk, shallot and ginger root, Tempe in coconut milk with shallot and ginger …). In the volume 12, Serat Centhini described that Tempe was made from soybean, in Bahasa Indonesia "kedelai," or in the Javanese language "kadhele": "*Kadhele Tempe srundengan, lombok kenceng lawan petis, gadhon rempah yem manjangan ....*" (Soybean Tempe with fried shredded coconut, shrimp paste with a lot of chili, spicy black eyed peas with string bean ….). Tempe has been a daily menu in Indonesian homes as a source of low-cost protein alternative to meat.

In 2010, more than three million tons of soybean have been consumed in a country of 237 million people, 975 thousand tons produced domestically, while the rest has been imported mostly from the USA and Argentina (Ministry of Trade 2011). Soybean is mainly used as raw material for producing Tempe and tofu with a production volume ratio of 50:50.

Tempe has bland taste and aroma, and high content of protein. It therefore is a flexible component in the diet and may be used as a simple meal or as ingredients to homemade recipes that home cooks could devise. For a quick daily meal, Tempe can be simply deep fried with a very low additional cost of frying palm oil, or transformed into crispy Tempe when fried in thin slices of 1 mm after being dipped into a light flour batter, or added to a vegetable coconut milk soup called *lodeh Tempe*. When Tempe was introduced to the western countries, it was easily incorporated into western cuisine in the form of products such as Tempe burgers, Tempe pizza, and Tempe sandwiches. In Japan, Tempe was incorporated into sushi which was appropriately called sushi Tempe.

13.1.2 From Home Industry Toward Modern Industry

Tempe, fermented by *Rhizopus* sp., was traditionally originally made manually at home and in cottage industries, using dried old Tempe as a starter, and banana leaves as the wrapper. To form Tempe, a threshold level of starter had to be present to reach such needed levels of *Rhizopus* sp., and the process in cottage industries had to go through several periods of fermentation cycles (3–4 cycles) before sufficient fermentation would have occurred to transform the soybean mix into Tempe. This occurred because the atmosphere in the facility had to provide a threshold level of mold spores for the process to progress well. Commercially available *Ragi* Tempe made from Rhizopus spp. spores is now used as the Tempe starter, and is widely available in Indonesia supplying hundreds of thousands Tempe cottage industries. When Tempe found its way to Europe, the USA, Japan, and Australia, the industry there became much more hygienic, with designated and specific designed fermentation rooms.

The last 15 years has seen rapid growth in the production of Tempe. In Indonesia, this snack food is manufactured and sold as Tempe chips, canned Tempe curry, and as Tempe powder used to manufacture various products such as weaning food and ice cream.

13.2 The Traditional Tempe Process

13.2.1 Tempe Processing in Cottage Industry

Various authors have described the highly varied Tempe processing methods in cottage industry in Indonesia (Hermana and Karmini 1996; Syarief et al. 1999; Shurtleff and Aoyagi 2001; Nouts 2001; Fung and Crozier-Dodson 2008). There are yet further practices in West Java that produce variants to what could be considered as mainstream Tempe production but are actually the norm in these particular localities (Fig. 13.1).

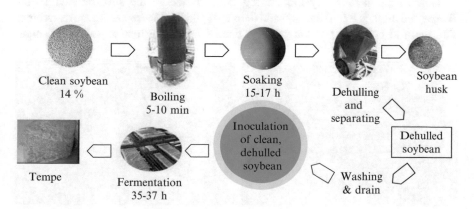

Fig. 13.1 Tempe processing in a traditional home industry

The manufacture of Tempe involves the use of dry and clean soybean with a moisture content around 14 % wet basis which is heated to boiling temperature and kept at so for about 5–10 min, and the whole process may last for 2–3 h depending on the capacity of the soybean cooking container. The boiled soybean is then soaked in cold water for about 15–17 h, drained, and dehulled using a burr mill. The soybean coat is removed, and the dehulled bean is washed, drained, and inoculated with commercial *ragi* Tempe (Tempe starter) with a ratio of 10 kg bean to 1–3 tablespoons of *ragi* depending on the room temperature. The higher room temperature, the smaller the quantity of *ragi* Tempe required for the inoculation. In a tropical country such as Indonesia, temperature in lowland areas is most suitable for the growth of *Rhizopus oligosporus*, i.e., 25–35 °C. After the *ragi* Tempe and the soybeans are mixed well, the mixture is transferred to a punctured plastic film bag to allow sufficient oxygen to enter. The mix is left to ferment for 35–37 h. The Tempe produced could be sent to market and presented inside the same bag. On the other hand, commercial Tempe from cottage industries may either be presented wrapped in banana leaves or in plastic film (Fig. 13.2).

13.2.2 Traditional Process of Tempe Starter Culture

In cottage industries, Tempe fermentation is commonly initiated by introducing Tempe starter culture made from the previous batch of Tempe produced. The traditional starter is available in two major formats: *laru* in powder form and *usar* in solid form. Commercially manufactured *Laru* Tempe is also known as *ragi* Tempe. *Laru* Tempe is produced from thin sliced Tempe fermented on round bamboo plates (tampah) under the cover of banana leaves and left for about 30–33 h to let the mold mycelium grow and produce spores. It is then dried under the sun, pounded with a stone mortar, and sieved with a bamboo or metal sieve to obtain fine powder. The fine powder is then mixed with roasted rice flour in a ratio of one *laru tempe* to ten rice flour (w/w), and packed in a film bag (Suliantari 1996).

Laru beras is made by inoculating cooked rice as the substrate with *laru Tempe*; the mold is left to develop and then the spores of *Rhizopus* sp. are harvested. The method is similar to the process used for the production of laru Tempe.

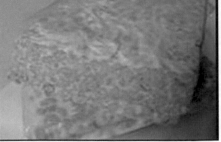

Fig. 13.2 Commercial Tempe from home industry wrapped in banana leaves and in plastic film

The introduction of cooked rice as substrate yields a higher volume of end product, with a shorter fermentation time to about 24 h. The starter thus produced is however regarded as having lesser purity. The production of *Laru singkong* involves the use of cooked cassava chips or moistened cassava chip powder as substrate for fermentation that lasts about 36 h.

Usar is produced from inoculated cooked dehulled soybean mixed with *laru Tempe*. A small amount of mixture, about one tablespoon, is spread on top of leaves with suitable surfaces such as waru (*Hibiscus tiliaceus*) and teakwood (*Tectona grandis* Linn. f.) leaves. The leaf structure of these plants is such that there is sufficient oxygen in the microenvironment for the fermentation process. The leaves are turned over and piled one on top of the other, and afterwards covered with banana leaves or cloth and left to undergo the first stage of fermentation that lasts about 24–36 h. In the second stage of fermentation, the leaves — now facing upwards — are separated individually, and kept so for 3–7 days, partly to let the *usar* dry at room temperature. Sun drying may be applied at the end of the second stage of fermentation to ensure that the *usar* reaches a sufficiently low moisture content that will secure a shelf life of about 7 days.

Laru and *usar* are traditional cottage industry products and as such tend not to be the product of pure inoculants. Studies on various *laru* and *usar* products from various locations in Java found out that *laru* and *usar* might contain various molds of Rhizopus sp., such as *R. oligosporus, R. oryzae, R. arrhizus,* and *R. stolonifer,* together with other molds, such as *Mucor rouxii, M. javanicus,* and *Fusarium* sp., and yeasts such as *Trichosporon pullulans* (Suliantari 1996).

13.2.3   Traditional Foods Made from Tempe

Traditional home recipes that transform Tempe into various meals are very varied in their composition and processing. Tempe as a bland raw material with high protein that lacks taste and its own aroma. It is therefore a good base component that can easily be blended into whatever meals that people might think of (Table 13.1 and Fig. 13.3). One is simple fried Tempe with no crust, the second is fried Tempe with crust, and the third is a vegetable soup with coconut milk and Tempe.

13.3   Industrialized Process of Tempe

13.3.1   Modern Processing of Tempe Industry

Development of tempe cottage industry into small-scale industry is mainly due to the introduction or upgrade of the dehulling machine, cooking equipment, the use of commercial *ragi* tempe as the source of *Rhizopus oligosporus*, the use of hygienic fermentation chambers, film packaging, and ceramic floosr and walls in

Table 13.1 Simple traditional foods made from Tempe

Recipe	Description
1. Deep-fried tempe	Slice 200 g of tempe into 0.5 cm thickness into rectangular shapes. Dip into a solution of 100 mL water mixed with one teaspoon of salt and one teaspoon of tamarind paste. Deep fry in any vegetable frying oil until the color on all sides is light brown. Drain and serve hot for 1–2 persons
2. Crispy tempe	Grind or pound in a stone mortar 1 clove of garlic, 1 candle nut, and 2 leaves of kafir lime. Put into a mixture of 100 mL water, 40 g of rice flour and 20 g of tapioca. Add about half tablespoon salt or as required to taste. Slice 200 g of tempe into very thin slices of 2 mm thickness. Dip the sliced tempe into the mixture until they are well coated by. Deep fry, serve for 1–2 persons
3. *Lodeh* tempe	Saute 5 shallots, 2 cloves of garlic, 1 chilli, 20 g crushed *lengkuas* (court case root), and 3 bay leaves. Add 200 mL milk made from half a coconut. Heat until the mixture is simmering, add 300 mL water then put in 200 g of cubed tempe (2 cm×2 cm), 130 g of sliced carrots, 70 g of string beans cut into 4 cm lengths, 80 g sliced baby corn, 30 g leaves of *Gnetum gnemon*, 1 corn cob sliced into 1.5 cm lengths, 100 g sliced *gambas* (vegetable pear). When the mixture boils add salt and sugar to taste. Serve hot for 4–5 persons

Fig. 13.3 Simple traditional foods made from Tempe: (**a**) deep-fried Tempe, (**b**) fresh sliced Tempe (**c**) crispy Tempe, and (**d**) fresh ingredients for *lodeh* Tempe, and (**e**) the soup ready for serving

the processing area. Modification and variation of the equipment mostly were in accordance with the needs of local conditions and the availability of technical support for the maintenance of equipment. Although the Tempe industry in Indonesia is considered as small-scale and cottage industry, and often regarded as a small village craft, the sheer number of individual factories however made these practice an overall huge industry (Shurtleff and Aoyagi 2001). Estimated data mentions that there are about 220 thousand small-scale and home Tempe industries all around Indonesia (KOPTI 2010). There is significant variation in the production capacity of the various production plants, such that a tempe cooperative might be capable of processing 2–5 tons soybean/day, while the production average of a home industry is around 150 kg soybean/day.

In general, the modernization of Tempe industries (Fig. 13.4) has retained the unique qualities of the traditional and cultural heritage of Tempe which historically always belonged to the common people. The modernization of the Tempe industry was more of an expansion process to increase the number and size of the rural factories. This change was brought about to match consumer needs rather than creating new generation of tempe products for domestic as well as export markets. Sutrisno (1999) theorized that the modernization of the Tempe industries had three targets: first, to increase Tempe consumption equally at all level of the society, and this could be part of a health strategy program from the government. Secondly, to develop a way to improve the existing Tempe home industry, so that they fulfil the food and hygienic standards befitting a modern food industry while raising productivity. Third, to improve the livelihood and welfare of all those working in Tempe industries. In general, the product of small-scale Tempe industry would conform to the Indonesian National Standards for Tempe that is periodically revised. One such revision was published in 2009 as SNI Tempe Kedelai (Soybean Tempe) No. 3144:2009.

13.3.2   The Commercial Tempe Inoculants

The preparation of Tempe inoculums in a laboratory scale has been discussed in details by Tanuwidjaja (1995), and Prawiroharsono (1999). The Modern Tempe starter industry took off as early as 1976, and contributed significantly to the national Tempe industry and economic development in general. The distribution of *ragi*

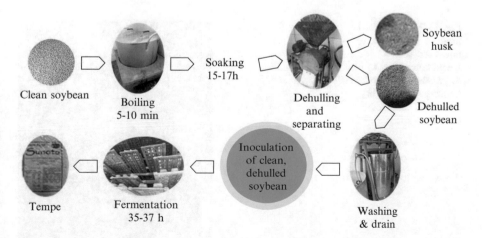

Fig. 13.4 Small-scale Tempe industry in Jakarta (Tempe Super Sunoto) using more hygienic equipment

tempe (literally translated tempe yeast), the market brand of the starter, has been long supported by the Indonesian Cooperatives of Tofu and Tempe (KOPTI) founded in 1977 and which has over than 220 thousand members around Indonesia.

The outline of the processing of tempe starter is steam cooking of rice as the basic material, cooling, mixing of cooked rice with previous inoculums in a ratio of 1 g inoculums to 1 kg of cooked rice, fermentation for 3 days until the cooked rice becomes overgrown with mold and spores, mechanical drying at 40 °C, milling into powder, and packaging into bottle or plastic film bag. The commercial product of Tempe starter or *ragi Tempe* made by the modern industry is presented in Fig. 13.5.

13.3.3 Tempe Snack Industry and Tempe Canning Industry

Starting early 1980s, small Tempe chips were developed as a snack and proved to be highly popular in Indonesia, mainly in Java. Tempe is sliced into various forms of chips such as rectangular, round, and perpendicular, and either deep fried or vacuum fried in vegetable oil. The fried Tempe is coated with various kinds of seasoning powder with flavor such as cheese, barbeque, shrimp, *balado* (hot and spicy), pizza, burger, spaghetti bolognese, sweet roasted corn, hot roasted corn, kebab, seaweed, black pepper steak, and meat ball. The packaging can be made from plastic film and aluminum foil and branded for marketing purposes (Fig. 13.6).

Tempe is also commonly incorporated into Indonesian meals prior to canning. Thus, the consumer has access to Tempe which is ready to eat when the can is reheated and opened. Tempe curry is one such favourite (Fig. 13.7) along with a new innovative product called "tempe qoir" which is a thick and sweet broth made from Tempe. This latter product was selected as one of the 100 innovations in Indonesia for 2010.

Fig. 13.5 Commercial product of *ragi* Tempe, powdered Tempe inoculums, in half kg film bag made by the modern industry (Courtesy of Research Centre for Chemistry-LIPI)

Fig. 13.6 Various kinds of Tempe chips in the market

Fig. 13.7 Tempe curry and qoir in can (Courtesy of Research Centre for Chemistry-LIPI)

13.4 Nutritional Value and Functional Properties of Tempe

13.4.1 Nutritional Value of Tempe

During relatively short fermentation, *R. oligosporus* is able to convert soybean nutritional components into various other substances, which gives Tempe specific new flavors, higher nutritional value, and digestibility. Fresh tempe contains about 64 % moisture, 18 % protein, 4 % fat, and 13 % carbohydrate (Hermana 1972). Astuti et al. (2000) and Fung and Crozier-Dodson (2008) reviewed various chemical and biochemical changes during the fermentation process of Tempe. One significant finding was that the soybean trypsin inhibitor which was undesirable to human health was produced by the protease released from the mold *Rhizopus oligosporus*. Further heat processing of Tempe would then inactivate the soybean trypsin. Free fatty acids were also produced through triacylglycerol hydrolysis caused by a lipase enzyme; however, these fatty acids were used by the mold itself, thus linoleic acid, palmitic acid, and stearic acid decreased in concentration during the life cycle of the mold.

Although the protein content of tempe does not differ much from that of unfermented soybean at the same moisture content, the water-soluble nitrogen content in tempe is significantly higher when compared to that of the unfermented soybean, approximately by a factor of six. Steinkraus et al. (1965) found that the water-soluble nitrogen content of dehydrated tempe was 39.0 %, while unfermented soybean contained only 6.5 %, at the same 2 % moisture. However, Astuti et al. (2000) reviewed that the solubility of nitrogen content increased from 3.5 mg/g in unfermented soybean to 8.7 mg/g in Tempe. Proteolytic enzymes in the mold during the fermentation process were responsible for breaking down higher molecular proteins in soybean to smaller nitrogen-containing molecules, including water-soluble amino acids. The pre-hydrolysis process on protein during the fermentation seems to promote better nutritional utilization by the human body, as a consequence of better digestibility of the product.

In general, protein quality measured as PER (protein efficiency ratio) of tempe remained relatively constant throughout the course of fermentation time. Hackler et al. (1964) reported that the PERs of soybean during 0, 24, 26, and 72 h of tempe fermentation were 2.63, 2.56, 2.49, and 2.44, respectively. While, Zamora and Veum (1979) reported an increase in the average of daily weight gain for weanling rats fed tempe compared with the daily weight gain for rats fed unfermented soybeans. It was also reported that tempe had a greater apparent biological value and net protein utilization compared with unfermented soybean.

Fung and Crozier-Dodson (2008) have stated that the amount of free amino acids such as tryptophan increased; however, long fermentation over 24 h might cause losses in threonine, lysine, and arginine. Further changes during Tempe fermentation were the increased levels of monosaccharides such as glucose that were produced as a consequence of enzymatic degradation while levels of disaccharides and the oligosaccharides were lowered. Thus, the decreased levels of raffinose and stachyose that commonly caused flatulence was another beneficial characteristics of Tempe.

In general, studies had agreed that even though the level of trace minerals did not change, the solubility of the minerals had been increased. This might be due to the reduction of 22 % phytic acid during fermentation which in turn would improve the bioavailability of the minerals (Sudarmadji and Markakis 1977). Increased solubility of iron, zinc, copper, and magnesium had been cited, but there was no clear indication whether potassium and calcium behaved in either way (Astuti et al. 2000; Fung and Crozier-Dodson 2008).

The production of vitamin B_{12} during Tempe fermentation has been long confirmed by various scientists. Keuth and Bisping (1994) concluded that *Klebsiella pneumoniae* or *Citrobacter freundii* were responsible in the formation of vitamin B_{12} in Tempe. The vitamin content of tempe was higher than the unfermented soybean in certain cases, but lower in others. Steinkraus et al. (1961) reported that riboflavin content was doubled in 1 g tempe (7 μg) compared with that in 1 g soybean (3 μg), niacin was approximately seven times in 1 g tempe (60 μg) compared with that in 1 g soybean (9 μg), and vitamin B_{12} was approximately increased 33 times in 1 g tempe (5.0 ng) compared with that in 1 g soybean (0.15 ng). Thiamine decreased from 10 μg in 1 g soybean to 4 μg in 1 g tempe, and also pantothenate decreased from

4.6 µg in 1 g soybean compared with 3.3 µg in 1 g tempe. Murata et al. (1970) found that biotin and total folate compounds in tempe were respectively 2.3 and 4–5 times higher than in unfermented soybean. The increase in riboflavin, niacin, vitamin B_{12}, biotin, and folate is very important nutrition changes in tempe fermentation, in particular vitamin B_{12}, which is usually comes from meat and milk.

Murakami et al. (1984) and Wuryani (1995) had concluded that Tempe contained isoflavone which was useful for human health, and found that the level of isoflavone was higher than in tofu and soy beverages.

13.4.2 Functional Properties of Tempe

Reduction in diarrheal symptoms was studied by Karmini (1987) by infecting rabbits with *E. coli* after feeding for 4 weeks. In 14 days observation, afterwards, it was found that the diarrheal symptom was indicated in 36 % rabbits fed with Tempe, and in 64 % rabbits not fed with Tempe. This study has supported a report by Dutch scientists observing the prisoners of war in the Japanese camp in Java during the Second World War (Shurtleff and Aoyagi 2001).

Astuti et al. (2000) reported that a Tempe diet could decrease cholesterol levels. After 3 month fed with an instant Tempe formula drink, the LDL cholesterol decreased by 12.0 % in male respondents and 9.7 % in female respondents. Cholesterol levels subsequently increased again when the feeding was stopped for 2 months.

The possibility of Tempe to help in menopausal symptoms and cancer prevention had been discussed, but the theory was based only on the lower incidence among Asian communities consuming high level of soy foods (Karyadi and Lukito 1997; Astuti et al. 2000). No specific studies have been conducted to confirm reports.

13.5 Present Development of Tempe Derived Products

13.5.1 Tempe Powder and Tempe Weaning Food

Fresh Tempe is a perishable food with a relatively short shelf life, therefore production and development of second generation of Tempe products for their versatility are should be explored. The second generation of Tempe products can be used to enrich high protein in food alternatives for people who suffer from nutritional deficiencies or to play a role as a main ingredient for a new product altogether. The end products are usually designed for middle to high income earners who wish to eat for leisure and convenience and are more sophisticated in their tastes as are the urban and city communities. Various second generation of Tempe products are commonly manufactured using innovative technology processing and include products such as Tempe milk, sausage, hamburger, meat like products, pepperoni, salami, yoghurt, ice cream, cheese, jam, noodle, bread, sauce, meal mixed product, snack, savoury products, vegetable broth, and savoury pasta.

Tempe can be processed into Tempe powder used as raw material for making other food products such as cake, pastry, extrusion products, instant drink formula, and weaning food. The process to produce Tempe powder mainly consists of drying and grinding. Fresh Tempe is cut into small pieces, and subjected to blanching at 104 °C for about 10 min prior to pulverizing and drying. The dried Tempe is then ground and screened through a 30 mesh sieve (Astuti 1995). Tempe powder can be utilized for making weaning food formula. The following is an example of a weaning food formula made from Tempe powder: 42 % roasted rice flour, 40 % precooked banana flour, and 18 % Tempe powder. Other alternative product for weaning food is Tempe porridge made from 16 % Tempe powder, added by wheat flour, milk powder, sugar, and essence. To increase its valuable ingredient, Tempe can be fortified with other nutritious materials such as seaweed, carrots, unripe papaya pulp, and okara.

13.5.2 Tempe Milk, Tempe Ice Cream, and Tempe Probiotic Drink

Tempe milk is one alternative used to extend the potential use of Tempe. Fresh Tempe is cut into cubes of 1.5 cm × 1.5 cm × 1.5 cm, steamed, and extracted by the addition of water with 1:2 (w/v) ratio to obtain Tempe milk filtrate. Agar (0.08 % w/v), skim milk 4 %, and sucrose 7 % (w/v) are added into the Tempe milk filtrate and heated at 90 °C for 5 min. Filtration and bottling are required before the mixture is pasteurized (90 °C; 5 min). The Tempe milk is finally ready to be consumed after cooling at 10 °C (Susanto et al. 1997). The process of Tempe ice cream is similar to the process of ice cream from milk. In this process, part of skim milk is substituted with Tempe powder, or fresh Tempe which is transformed into Tempe paste by blanching and grinding. Water and other ingredients such as sucrose, fat, emulsifier, flavor, and food color are evenly added, mixed, and homogenized before being pasteurized. Then, the mixture is frozen into ice cream and packed before distribution.

Tempe is also claimed to have active estrogen like isoflavone compound which is useful to menopause women. The use of Tempe as basic ingredient for making probiotic drink or yoghurt inoculated by *Lactobacillus bulgaricus* and *Streptococcus thermophilus* is beneficial to the extent of providing low-cost food supplement for the menopause women. The Tempe probiotic drink is produced by mixing and blending fresh milk with Tempe in a ratio of 2:1 (v/w). The mixture is then added by 50 % whey protein (v/v), pasteurized, and inoculated by the starter culture. The probiotic drink can also be transformed into powder by the addition of 10 % dextrin, and dried at 50 °C for 8 h, ground, and sieved. Analysis of the probiotic drink powder indicates the isoflavone content of 75 mg/100 mg, and the number of lactic acid bacteria of 10^6 cell/g (Kumalaningsih and Padaga 2005).

13.5.3   Tempe Sausage

Tempe has also been developed to make sausages. The fresh Tempe was cut into small pieces and ground. A mixture containing egg white, wheat flour (1:1), water, garlic, and spices is prepared, and the ground Tempe then is poured into the mixture. Forty gram of vegetable oil is added followed by continuous mixing yielded to the final mixture. The final mixture then is filled into the sausage casing prior to steaming at 100 °C (Susanto et al. 1997).

13.6   Research and Commercial Development in Other Countries

A scientific article about tempe was first published by a Dutch researcher H.C. Prinsen Geerligs in 1895 (Shurtleff and Aoyagi 2001), during the Dutch occupation era on Indonesia for 300 years from 1642 to 1942. Geerligs and his colleagues further studied the microorganism and were of the opinion that *Rhizopus oryzae* was the mold grew in tempe. It was thought that there may have been some confusion with anther Indonesian traditional fermented food called *oncom* which was made from peanut cake, the by-product of peanut oil, inoculated with this mold in a similar pattern as Tempe. The first Japanese scientists to publish report on Tempe were believed to be Ryoji Nakazawa and his co-worker Yoshito Takeda in 1928 (Shurtleff and Aoyagi 2004). The research on Tempe in Japan then continued until at present, driving the development of Tempe industry in the country both by Indonesian migrants and Japanese community.

In 1935, I.H. Burkill wrote a dictionary about economic products in Malay Peninsula covering some information about Tempe (Shurtleff and Aoyagi 2001). The first journal article was published by Gerald Stahel in 1946, as a result of study in Suriname (a Dutch colony) where many Javanese migrants had settled. In 1954, American scientists took up research on Tempe that had been started by Paul György, and Kiku Murata. They were followed by Keith Steinkrauss an American scientist assisted by Yap Bwee Hwa from Indonesia, who published a journal article in 1960 that was regarded as a milestone in Tempe research history. A significant contribution was made by C.W. Hesseltine and Ko Swan Djien from Indonesia in 1963 who found that *Rhizopus oligosporus* was the mold growing on tempe, and later on recommended the use of film packaging for tempe as an alternative to banana leaves wrapping and which eventually became widely used together with the traditionally banana leaves. Many Indonesian researchers continued to study tempe abroad such as Slamet Sudarmadji with Pericles Markakis from the USA who investigated the chemical and nutritional aspects of tempe in 1973–1977 (Sudarmadji and Markakis 1977), and Ko Swan Djien with research scientists from the Agricultural University, Wageningen, the Netherlands who characterized the production of tempe inoculums (Rusmin and Ko 1974). In 1996, many Indonesian researchers together with the worldwide scientists have written scientific articles about Tempe in a handbook of indigenous fermented foods (Andersson et al. 1996).

Commercially manufactured Tempe has been produced and sold widely around the world from Europe, Asia, the USA, to Australia, and later to Africa, Asia Pacific, and Latin America. The Tempe burger and Tempe made from five different grains were originally produced in the USA in 1980s (Tibbott 2004). The latter development has indicated that organic Tempe has increasingly attracted consumers such as Yakso Tempeh made in the Netherlands (Healthy Supplies 2015) and Natursoy Tempeh made in Spain (Natursoy 2015) which are packed in glass bottles, Marusa Tempeh made in Japan and packed in vacuum film packaging (Marusa Tempeh 2015), and Rhapsody Organic Tempeh made in the USA packed in film (Rhapsody Natural Foods 2015).

Tempe has found its way into western food recipes in the form of Tempe pizza, Tempe burger, Tempe sandwiches, Tempe barbeque, and to Japanese cuisine as well as sushi Tempe. In 1980s, a Californian snack industry manufactured freeze-dried Tempe in a cardboard tubular packaging. Compared to the Indonesian Tempe industry, Tempe industries in developed countries have been significantly modernized to include fully mechanized equipment, separate fermentation rooms with other processing areas, and improved hygienic design of the facilities that included appropriate drainage, sanitary floors and walls, and waste treatment.

Shurtleff and Aoyagi (2001) listed eight major tempe industries with a capacity from 2 to 7 tons soybean per week, one located in the Netherlands, three in Japan, and four in the USA. Listing of Tempe manufacturers worldwide is available in the website (Tempe Manufacturers 2015). The list shows that the availability of this product is constantly increasing globally; this augers well for the consumption of what started as a homemade soya bean product which has now become a global industry.

References

Andersson FE et al (1996) Indonesian tempe and related fermentations: protein-rich vegetarian meat substitutes. In: Steinkraus KH (ed) Handbook of indigenous fermented foods, 2nd edn. Marcel Dekker, New York, pp 1–94

Astuti M (1995) Perkembangan industri tempe di Indonesia. In: Prosiding Simposium pengembangan industri makanan dari kedelai, LIPI, Jakarta, 23 Sept 1995

Astuti M (1996) Sejarah perkembangan tempe. In: Sapuan, Sutrisno N (eds) Bunga rampai tempe Indonesia. Yayasan Tempe Indonesia, Jakarta, pp 21–42

Astuti M et al (2000) Tempe, a nutritious and healthy food from Indonesia. Asia Pac J Clin Nutr 9(4):322–325

Fung DYC, Crozier-Dodson BA (2008) Tempeh a mold-modified indigenous fermented food. In: Farnworth ER (ed) Handbook of fermented functional foods. CRC, Boca Raton, pp 476–493

Hackler HR et al (1964) Studies on the utilization of tempeh protein by weaning rats. J Nutr 82(4):452–456

Healthy Supplies (2015) Organic tempeh. http://www.healthysupplies.co.uk/tempeh-fermented-soya-yakso.html. Accessed 19 July 2015

Hermana (1972) Tempe- an Indonesian fermented soybean food. In: Stanton WR (ed) Waste recovery by microorganisms. UNESCO/ICRO-University of Malaya, Kuala Lumpur, pp 55–62
Hermana, Karmini M (1996) Pengembangan teknologi pembuatan tempe. In: Sapuan, Sutrisno N (eds) Bunga rampai tempe Indonesia. Yayasan Tempe Indonesia, Jakarta, pp 151–168
Karmini M (1987) The effect of tempe on the control of enteropathogenic diarrhea. Dissertation, Bogor Agricultural University
Karyadi D, Lukito W(1997) Functional characteristics of tempe in disease prevention and treatment. In: Sudarmadji, Suparmo, Raharjo (eds) Reinventing the hidden miracle of tempe. Proceedings international tempe symposium, Bali 13–15 July 1997. Indonesian Tempe Foundation, Jakarta, pp 199–204
Keuth S, Bisping B (1994) Vitamin B12 production by Citrobacter freundii or Klebsiella pneumoniae during tempeh fermentation and proof of enterotoxin absence by PCR. Appl Environ Microbiol 60(5):1495–1499
KOPTI (2010) Laporan Tahunan. KOPTI, Jakarta
Kumalaningsih S, Pradaga M (2005) Formulation and development of finished products of functional foods: foods supplement for menopause women. In: Karossi AT (ed) ASEAN training on functional foods: trend and challenges. LIPI, Jakarta, pp 79–88
Marusa tempeh (2015) Tempeh. http://www.izu-marusa.com/. Accessed 20 July 2015
Ministry of Trade (2011) Soybean import statistics. Ministry of Trade Republic Indonesia, Jakarta
Murakami H et al (1984) Antioxidative stability of tempeh and liberation of isoflavones by fermentation. Agric Biol Chem 48(12):2971–2975
Murata K et al (1970) Studies on the nutritional value of tempeh III. Changes in biotin and folic acid contents during tempeh fermentation. J Vitaminol 16(4):281–284
Natursoy (2015) Productos tempeh. https://www.natursoy.com/productos-ecologicos//tempeh/tempeh. Accessed 19 July 2015
Nout MJR (2001) Fermented foods and their production. In: Adams MR, Nout MJR (eds) Fermentation and food safety. Aspen, Gaithersburg, pp 1–38
Prawiroharsono S (1999) Microbiological aspects of tempe. In: Agranoff J, Sapuan, Sutrisno N (eds) The complete handbook of tempe: the unique fermented soyfood of Indonesia. American Soybean Association, Singapore, pp 93–97
Rhapsody Natural Foods (2015) Organic tempeh. http://rhapsodynaturalfoods.com/our-products/tempeh/organic-tempeh/. Accessed 19 July 2015
Rusmin S, Ko SD (1974) Rice-grown Rhizopus oligosporus inoculum for tempeh. Appl Microbiol 28(3):347–350
Shurtleff W, Aoyagi A (2001) The book of tempeh, 2nd edn. Harper & Row, New York
Shurtleff W, Aoyagi A (2004) History of tempeh. In: Shurtleff W, Aoyagi A (eds) History of soybeans and soyfoods, 1100 B.C. to the 1980s. Unpublished manuscript. http://www.soyinfocenter.com/HSS/tempeh4.phpp. Accessed 19 July 2015
Steinkraus KH et al (1961) Pilot plant studies on tempe. In: Proceedings Conference on soybean products for protein in human foods, USDA ARS 72-22, Peoria, 13–15 Sept 1961
Steinkraus KH et al (1965) A pilot process for the production of dehydrated tempeh. Food Technol 19(1):63–68
Sudarmadji S, Markakis PP (1977) The phytate and phytase of soybean tempeh. J Sci Food Agric 28(4):381–383
Suliantari (1996) Pembuatan dan penanganan laru (inokulum) tempe. In: Syarief R (ed) Pengembangan industri kecil tempe. Kantor Menneg Urusan Pangan-IPB, Jakarta, pp 2.1–2.24
Susanto T et al (1997) Research on the utilization of tempe as raw material in the production of milk and tempe sausage. In: Sudarmadji, Suparmo, Raharjo (eds) Reinventing the hidden miracle of tempe. Proceedings international tempe symposium, Bali 13–15 July 1997. Indonesian Tempe Foundation, Jakarta, pp 125–130
Sutrisno N (1999) The politics of developing national tempe industry. In: Agranoff J, Sapuan, Sutrisno N (eds) The complete handbook of tempe: the unique fermented soyfood of Indonesia. American Soybean Association, Singapore, pp 166–170

Syarief R et al (1999) Wacana tempe Indonesia. Universitas Katolik Widya Mandala, Surabaya, pp 50–65

Tanuwidjaja L (1995) Perkembangan industri tempe di Indonesia. In: Prosiding Simposium pengembangan industri makanan dari kedelai, LIPI, Jakarta, 23 Sept 1995

Tempe Manufacturers (2015) Tempeh4u. http://tempeh4u.myweb.hinet.net/entempehmakers. html. Accessed 19 July 2015

Tibbott S (2004) Tempeh: the other white beancake. In: Hui YH et al (eds) Handbook of food and beverage fermentation technology. Marcel Dekker, New York, pp 583–594

Wuryani W (1995) Isoflavone in tempe. ASEAN Food J 10(1):99–102

Zamora RG, Veum TL (1979) The nutritive value of dehulled soybeans fermented with Aspergillus oryzae or Rhizopus oligosporus as evaluated by rats. J Nutr 109(7):1333–1339

Chapter 14
Modernization of Manufacturing Process for Traditional Indian Dairy Products

P.S. Minz and R.R.B. Singh

Contents

14.1 Introduction

A wide variety of traditional dairy delicacies, drawn from different regions of the country, are produced in India using processes such as heat and/or acid coagulation, desiccation and fermentation. These products play a significant role in the economic, social, religious and nutritional functions of the Indian masses. The total milk production in India for 2013–2014 was 137.7 million tonnes (DAHDF 2015). It is estimated that about 50–55 % of milk produced is converted by the traditional sector into variety of Indian milk products (Patil 2005). The increased availability

P.S. Minz
National Dairy Research Institute, Karnal 132 001, India

R.R.B. Singh (✉)
National Dairy Research Institute, Karnal 132 001, India

Sanjay Gandhi Institute of Dairy Technology,
BVC Campus, Jagdeo Path, Patna, Bihar 800 014, India
e-mail: rrb_ndri@rediffmail.com

© Springer Science+Business Media New York 2016
A. McElhatton, M.M. El Idrissi (eds.), *Modernization of Traditional Food Processes and Products*, Integrating Food Science and Engineering Knowledge Into the Food Chain 11, DOI 10.1007/978-1-4899-7671-0_14

of milk during the flush season coupled with the need to preserve it for long-distance transportation and marketing makes conversion of milk into traditional products particularly an attractive commercial proposition and great employment opportunities in the countryside. These traditional dairy products are classified based on the principles involved in their manufacture and could be an intermediate base for subsequent conversion into a wide varieties of delicacies or could be consumed as such in the form of the finished products (Table 14.1). The traditional methods of manufacture of these products have evolved over a long period and quality often depends on the skills of homemakers or local confectioners (*Halwais*). They generally use the batch methods for product preparation which results in higher processing cost and reduced energy efficiency. Due to production in small quantities, the local confectioners are limited to localized marketing. The chemical composition and organoleptic properties of traditional milk products also vary significantly depending on the skills of the workforce involved in its manufacture. With globalization of dairy trade and focus of the local industry on quality and consumer satisfaction, many large industries have taken to mechanized manufacture of these products on large scale. However, mechanization of the manufacturing process of these traditional dairy products is a very challenging task as the traditional processes involved in their manufacture are tailored to facilitate development of unique flavour and texture attributes in the product. Simulating these quality attributes in a product delivered by the mechanized process requires unique design considerations and manipulation of technological parameters and residence times. Systematic efforts have been made through novel approaches to either develop new mechanized equipment or adapt existing equipments for developing a continuous line for the manufacture of many traditional Indian milk products (Sharma et al. 2003).

14.2 The Products

14.2.1 Khoa

Khoa refers to an intermediate milk product traditionally obtained by desiccating milk over gentle heat until a thick viscous concentrate is formed. It serves as a base material for some of the most popular varieties of sweetmeats such as burfi, peda, gulab jamun, milk cake and kalakand. The Prevention of Food Adulteration Act (PFA) recommends that a good quality khoa must confirm to not less than 30 % milk fat on dry weight basis. To meet this regulatory norm, the minimum level of fat desired in buffalo milk should be 5.5 % while in cow milk it should be 4 % (Pal and Raju 2007). Generally, buffalo milk is preferred for the manufacture of khoa as it has larger proportion of butyric acid-containing triglycerides, and there is more release of free fat responsible for smooth and mellow texture which are desirable attributes in khoa (Sindhu 1996). Khoa made from cow milk lacks these characteristics and has moist surface, salty taste and sticky and sandy texture which are considered undesirable for the preparation of sweetmeats.

Table 14.1 Overview of classification of major Indian dairy products

Process	Product	Product type	Similar western product	Description
Heat desiccation	*Khoa*	Intermediate	Evaporated milk	*Khoa* has a dough like consistency and is prepared by continuous boiling of milk until desired concentration of solids (65–72 % TS) and texture are attained. It is used as a base material for a variety of popular sweetmeats, such as *burfi, peda,* and *gulab jamun*
Heat acid coagulation	*Chhana*	Intermediate	Lactic coagulated cheese	*Channa* is obtained by heating milk followed by cooling and acidifying it with suitable organic acid. It is the base material for the preparation of *rasogulla, sandesh, rasomalai,* etc. *Channa* with 50–60 % moisture content is preferred for sweetmeat preparation
	Paneer	Intermediate and Final	Soft cheese	*Paneer* refers to an indigenous variety of acid coagulated soft cheese. It forms base for a variety of culinary dishes, stuffing material for various vegetable dishes, snacks and sweetmeats
Fermentation	*Shrikhand*	Final	Sweetened quarg	Traditional fermented and sweetened milk product of Indian origin. It is prepared from solids (*Chakka*) recovered by draining the whey from lactic fermented curd. Sugar, cream and other ingredients like fruit pulp, flavour and colour are blended with *Chakka* to get semi-solid consistency
Heat clarification	*Ghee*	Final	Butter oil	*Ghee* is heat clarified butter fat for table use or as a frying medium. Apart from dietary usage, ghee is also used for performing religious rites

(a) *Traditional Method*: Buffalo milk (4–6 L) is boiled over direct fire in a shallow pan (mild steel) with vigorous stirring and scrapping. Within 5–10 min, a semi-solid mass having dough consistency is formed (Punjrath 1991).

(b) *Improved Methods*: *Khoa* manufactured by traditional method suffers from poor and inconsistent quality of the product. Attempts have been made to mechanize the production process of *khoa* using both the batch and continuous type plants (Aneja et al. 2002). These advances for commercial production of *Khoa* have been summarized in Table 14.2. Rajorhia et al. (1991) made a comparative study of quality of *khoa* prepared from different mechanical systems such as inclined scraped surface heat exchanger (ISSHE), conical vat, convap–contherm and roller dryer. Quality score was highest for ISSHE followed by roller dryer, conical vat and convap–contherm process.

Khoa has a great market demand but due to its short shelf life (5 days at 30 °C) it cannot be marketed over long distances. In the absence of proper packaging, the loss of moisture from the product adversely influences the texture and enhances the rate of chemical deterioration such as oxidation and browning. With the use of four-ply laminated pouches and tin containers, the shelf life of *khoa* can be increased up to 13 days at 30 °C and 75 days under refrigerated storage. Sterilization of packaging such as Polypack™ (Pitram Pura, Delhi, India) with gamma radiation using CO^{60} prior to product filling proved to be beneficial. Addition of 0.3 % potassium sorbate at the last stage of *khoa*-making increased the shelf life of *khoa* by another 10 days. Vacuum packaging of *khoa* could enhance the shelf life up to 120 days under refrigerated storage (Rajorhia et al. 1984).

14.2.2 Chhana

Chhana is a heat acid coagulated product having marble white colour, spongy texture with mild acidic flavour. It is used as a base material for manufacturing a large variety of sweets such as *Rasogolla, Sandesh, Rasomalai, chum chum* and *chhana murki*. Cow milk is preferred for manufacturing *chhana* as the product obtained is soft with smooth texture and velvety body which are highly desirable attributes for making *chhana*-based sweetmeats particularly *rasogolla*. Buffalo milk *channa* is generally hard and greasy because of inherent differences in qualitative and quantitative aspects of buffalo milk. However, technological interventions have been successfully employed to overcome these defects.

1. *Traditional Method*: Traditionally, cow milk is taken in a pan (2–40 L/batch) and is coagulated at high temperature (70 °C) using sour whey, but some organic acids such as citric acid, lactic acid and calcium lactate may also be used (Mathur 1991; Das 2000). Whey is then drained by straining through a muslin cloth. But in this method the yield is low due to draining of whey protein along with whey.

Table 14.2 Developments in mechanization of process for *khoa*-making in chronological order

Sl. No	Work done/ reported by	Description	Type of operation
1.	Banerjee et al. (1968)	The pilot plant has a scraped surface heat exchanger (SSHE) and two open semi-jacketed pans with reciprocating spring loaded scraper (Fig. 14.3). Milk with 12–13 % total solids (TS) is pumped into SSHE for concentration to 30–35 % TS. The first stage of the open semi-jacketed pan further concentrates the milk to 50–55 % TS. The final concentration to 70–75 % TS is achieved in the second pan. The equipment has a capacity of 50 L of milk per hour	Semi-continuous
2.	More (1985)	SSHE consists of a semi-jacketed drum with vapour exhaust. The scraper assembly comprises of central shaft and spring loaded blades with rubber boots (Fig. 14.4)	Semi-continuous
3.	Agrawala et al. (1987)	Conical process vat consists of stainless steel conical vat with cone angle of 60° with steam jacket partitioned into four segments for efficient use of thermal energy and less heat loss (Fig. 14.5). The mechanism consists of 3-equidistant arms supported at two points in the shaft and each arm having three independent spring loaded blades for scraping. A positive displacement screw pump is connected at the outlet of the vat for recirculation and spreading of the product over heat transfer surface	Batch
4.	Christie and Shah (1990)	It consists of a single stage SSHE, with silts on the top with hopper to collect foam during *khoa* manufacturing (Fig. 14.6). The steam jacket is provided at the lower part of the SSHE. Spring loaded scraper blades help in uniform milk heating and spreading. The equipment has a capacity of 50 L of milk per hour	Batch
5.	Punjrath et al. (1990)	Inclined scraped surface heat exchanger (ISSHE): In this machine, concentrated milk of 42–45 % total solids is used as feed. The inclination of ISSHE permits formation of a pool of boiling milk critical to formation of *Khoa* (Fig. 14.7). By varying the total solids, temperature and flow rate of feed, scraper speed and angle of inclination, product characteristics can be varied to meet the functional requirements	Continuous
6.	Patel (1991)	Double jacketed kettles. Heating is done through steam	Batch
7.	Verma and Dodeja (2000)	Two SSHEs are arranged in a cascade fashion (Fig. 14.8). Milk is concentrated into first SSHE to about 40–45 % TS and finally to *Khoa* in the second SSHE	Continuous
8.	Rajorhia (1995)	Roller drier can be used for preparing *khoa* by adjusting process variables such as steam pressure, roller speed, concentration and flow rate of milk and by changing the distance between the rollers and scrapper blades. Vacuum concentrated milk with 50 % total solids preheated to 74 °C for 10 min was found suitable for *khoa*-making on roller driers at 25–30 psi. A kneader is placed at the outlet of roller drier to make homogenous mass of *khoa*	Continuous
9.	Alfa laval Convap–contherm process (Aneja et al. 2002)	Concentrated milk from SSHE is pumped through a holding tube to impart desirable texture and flavour	Continuous

2. *Improved Methods*:

(a) *Ultrafiltration Process*: Application of Ultrafiltration (UF) process for *Chhana* manufacture resulted in 18–19 % extra yield due to higher recovery of whey proteins. In this process, heat-treated (92 °C for 5 min) skim milk is subjected to ultrafiltration followed by diafiltration (23 % TS) and the resultant retentate is mixed with plastic cream. The mixture is heated to 90 °C for 5 min and coagulated with lactic acid to develop soft coagulum. The granular mass is pressed to remove free moisture, yielding *Chhana* (Sharma and Reuter 1991).

(b) *Continuous Method*: It involves following steps (Singh 1994):

- Indirect heating of milk in a tubular heat exchanger to 95 °C
- Cooling to 70 °C
- Continuous coagulation with hot citric acid (70 °C) in a vertical tube
- Holding milk–acid mixture to permit complete coagulation
- Separation of whey in a continuous flow employing double wall basket centrifuge
- Chilling to 4 °C by directly spraying chilled water on the layer of *Chhana*

(c) *Casein Process*: Continuous casein making equipment can be adopted for large-scale production of *chhana* (Rajorhia 1995). Certain modifications of equipment required are:

- Intake of milk at 70 °C instead of 35 °C
- Strength of coagulant and milk to obtain a pH of 5.1
- Adequate residence time to effect co-precipitation of casein and whey proteins together with fat
- Mechanical removal of whey using a basket centrifuge or provision of additional vibrating screen as *chhana* drains too slowly
- Washing of coagulum once with potable water followed by pressing to retain about 60 % moisture in the finished product
- Arrangement for bulk packaging to synchronize with product delivery

14.2.3 Paneer

Paneer represents one of the semisoft varieties of the Queso-blanco type of cheese having a high moisture content of 50–60 % (Pal 2002). It is obtained by heat and acid coagulation of milk. It is consumed either in raw form or used in preparation of several varieties of culinary dishes and snacks. Good quality *paneer* has a characteristic white colour, sweetish, mildly acidic, nutty flavour, spongy body and close knit texture. Buffalo milk is considered more desirable for *paneer* manufacture as the product obtained has all these attributes. Buffalo milk also offers higher *paneer* yield due to higher total solids in it. Higher concentration of casein in the micelle state with bigger size, harder milk fat due to larger proportion of high melting triglycerides in it and higher content of total and colloidal calcium are responsible for producing desired quality of *paneer*. On the contrary, cow milk paneer has very compact and fragile body and its pieces may get disintegrated and loose their identity during cooking. The yield of *paneer* made from cow milk is also low as compared with buffalo milk (Pal and Raju 2007).

1. *Traditional Method*: Buffalo milk is standardized to 5.5 % fat and heated to around 90 °C. After coagulation with citric acid at 70–85 °C, a coagulum of casein–whey protein complex with entrapped fat is formed. Free whey is drained and coagulated mass is pressed in cloth-lined hoops. The coagulum knits together into a compact spongy mass under pressure. After removing from the hoops, blocks are placed in chilled water for firming. Blocks are cut into smaller pieces and are loosely placed in polythene bags for retail sale (Mathur 1991).

2. *Improved Methods*:

    (a) *Ultrafiltration Process*: Ultrafiltration (UF) when employed for *paneer* manufacture offers advantages of mechanization, uniform quality, improved shelf life, increased yield and a nutritionally better product. The process involves standardization and heating of milk followed by UF whereby lactose, water and some minerals are removed (Sachdeva et al. 1993). UF of milk and the removal of permeate are synonymous to removal of whey by coagulation in conventional method. The concentrated mass, which has about 40 % total solids, is cold acidified to get the desired pH. Texturization is achieved in a continuous process by using microwave tunnels. Such tunnels comprise of a series of magnetrons under which the product moves continuously on the conveyor belts. The resulting product has typical characteristics of normal *paneer* (Pal 2005).

    (b) *Nanofiltration Technique*: *Paneer* prepared from normal cow milk has hard, compact and dry characteristics. Nanofiltration of cow's milk helped to minimize these defects and produced better quality *paneer*, but imparted excessive brittleness. *Paneer* prepared from nanofiltered milk has higher moisture retention resulting in higher yield (Pal et al. 2002).

    (c) *Centrifugal Method*: After coagulation of milk (at 70 °C) using citric acid, primary whey is drained using gravity separation through muslin cloth. The coagulum is subjected to centrifugal pressing by double wall basket centrifuge at 30–60 °C to remove the residual whey. The pressed coagulum is chilled inside the basket centrifuge by chilled water at 4 °C for better body and textural characteristics. Pressing and chilling of coagulum by centrifugal method considerably reduces time for production of *paneer* (Agarwal 1996; Das 2000).

    (d) *Continuous Method*: Impact type device can compress blocks of coagulum to form *paneer* which could be taken out at regular intervals. For the pressing, coagulum is kept in cages made from a special type of screen and the cages are subjected to impact forces (Das and Das 2009).

(e) *Processed Paneer*: Processed *paneer* production is an attempt primarily aimed at evolving a product, which in its physico-chemical, sensory and functional characteristics has as close a resemblance as possible with the processed cheese. It has a lower cost relative to the western processed natural cheeses and can offer a variety in flavour, consistency and functionality. For manufacturing processed paneer, NaCl 0.5 %, emulsifying salt and flavour is mixed with shredded or comminuted paneer at 80 °C for 5 min. The mix is filled into moulds and cooled. Depending upon the type and intensity of the flavour desired, various dairy and non-dairy additives can be added (Pal 2002).

14.2.4   Shrikhand

1. *Shrikhand* is a very popular traditional fermented and sweetened milk product of India. It is prepared by lactic acid coagulation of milk and expulsion of whey from the curd followed by blending with cream, sugar, flavour and spices. *Shrikhand* has a typical semi-solid consistency showing characteristics firmness and pliability contributing to its suitability for consumption with "*Puree*" or "*Chapati*". The consistency is influenced to a great extent by the moisture, fat and sugar levels. Its colour varies from yellowish white to marble white depending on the type of milk used. *Shrikhand* has been traditionally prepared by home-makers or on small scale by the confectioners (*Halwais*). According to Prevention of Food Adulteration Act (PFA) of India, *shrikhand* shall contain not less than 8.5 % milk fat (on dry basis), not less than 9.0 % protein (on dry basis), and should not contain more than 72.5 % sugar (on dry basis). Same PFA standards imply to fruit *shrikhand* as for plain *shrikhand* except that the fat content in it shall not be less than 7.0 %. Buffalo milk is the preferred raw material for *shrikhand* as it has a high fat, SNF and Ca++ content. Fermentation of milk is accomplished by back-slopping, which contains mixed type of lactic acid bacteria. Higher fat content in milk is preferred as it produces a balanced delicate and pleasant flavour as well as desirable body and texture. *Chakka* (Quarg-like product) which serves as the base material is obtained by removal of whey from dahi (Yoghurt-like product). The quality of shrikhand depends significantly on the physical and chemical properties of *chakka*. Yield of *chakka* depends upon the heat treatment and total solids content of skim milk and starter culture (Aneja et al. 2002). The heating of milk to 85–90 °C for 16 s reportedly resulted in 24 % and 25 % yield of *chakka*, respectively. *Chakka* made from whole milk gives smooth body but leads to higher fat loss in whey, whereas skim milk *chakka* tends to be rough and dry.

2. *Traditional Method*: In the traditional method, buffalo or mixed cow milk is boiled and after cooling to room temperature (30–35 °C), it is inoculated with lactic culture and incubated for 6–8 h. When the curd is firmly set (acidity 0.9–1.0 %), it is placed in a muslin cloth bag and hung on a peg for drainage of whey for 6–8 h. The intermediate product (*chakka*) is used for the preparation of

shrikhand. The *chakka* thus obtained is mixed with sugar, kneaded well and rubbed through a muslin cloth to give a smooth product. Colours, flavours and fruits are added to provide variety (Sharma 1998; Punjrath 1991).

3. *Improved Methods*:

  (a) *Industrial Process*: In this process, skim milk curd is centrifuged in a continuous quarg separator to produce *chakka*. It is mixed with cream, sugar and flavourings in a scraped surface heat exchanger for manufacture and pasteurization of *shrikhand*. The product is filled in preformed cups under semi-aseptic environment before retail trade. The process line developed by National Dairy Development Board (NDDB) for *shrikhand* production is shown in Fig. 14.1.

  (b) *Ultrafiltration Technique*: Skim milk is heated in a double jacketed vat with slow agitation. It is cooled to 21–22 °C and inoculated with mixed starter culture (i.e. *Streptococcus lactic, S. cremoris* and *S. diacetylactis*) at the rate of 0.1–0.15 %. Incubation time varies from 16 to 18 h at 21–22 °C so as to get curd pH of 4.6–4.5 and a pleasant diacetyl aroma. Rapid fermentation may alternatively be done with the help of yoghurt culture (*Streptococcus thermophilus* and *Lactobacillus bulgaricus*) requiring 4 h of incubation. After ultrafiltration, the retentate is pressed to get *chakka*. *Chakka*, cream (70 % fat) and sugar are mixed in a planetary mixer (30–35 rpm) for half an hour to get a product with smooth texture, plastic body and a sweet-acidic flavour (Sharma and Reuter 1992; Sharma 1998). Addition of colour and flavour is optional. The process flow of *shrikhand* manufacture using ultrafiltration (UF) is shown in Fig. 14.2. The yield is 23.16 % higher in UF process as compared to traditional process because of better recovery of whey proteins.

  (c) *Reverse Osmosis (RO) Process*: This process involves heating of milk (90 °C for 5 min), application of RO, cooling to 22 °C, inoculation with 20 % mixed lactic culture, incubation for 18 h and then removal of whey by filtration to get *chakka* (Sachdeva et al. 1994). Increased yield, higher solid recovery, reduced processing time, access to mechanization and alleviation of whey disposal problem are the major advantages of the process.

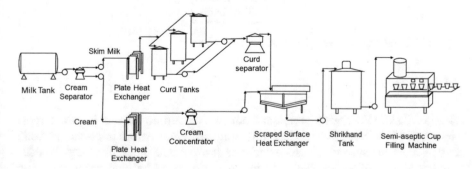

Fig. 14.1 Industrial method for Shrikhand production

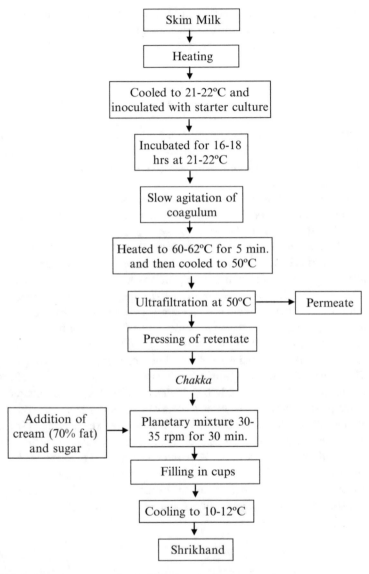

Fig. 14.2 Ultrafiltration process for Shrikhand production

14.2.5 *Ghee*

Ghee is clarified butter fat prepared from cow or buffalo milk. It has the largest share in trade of traditional Indian milk products. *Ghee* has an important place in Indian diets because of its characteristics flavour and pleasant aroma, besides being a source of fat-soluble vitamins. It is produced by heat desiccation and clarification of *makkhan*, butter or cream at 105–110 °C. *Makkhan* here refers to

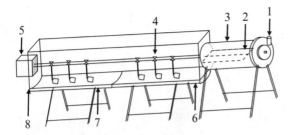

1	Milk inlet
2	Rotary scrapper
3	SSHE
4	Reciprocating scraper
5	Driving mechanism
6	Steam jacketed open pan 1
7	Steam jacketed open pan 2
8	*Khoa* outlet

Fig. 14.3 Pilot plant for *Khoa*-making

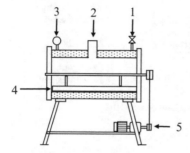

1 Steam inlet
2 Milk inlet
3 Pressure gauge
4 Scraper
5 Driving mechanism

Fig. 14.4 Semi-continuous *Khoa*-making plant

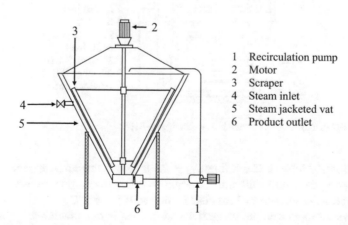

1	Recirculation pump
2	Motor
3	Scraper
4	Steam inlet
5	Steam jacketed vat
6	Product outlet

Fig. 14.5 Conical process vat

the country butter normally obtained by churning whole milk curd with crude indigenous devices. Heat-induced changes in milk protein/lactose during the clarification process impart a distinctive, pleasant cooked flavour to ghee (Aneja et al. 2002).

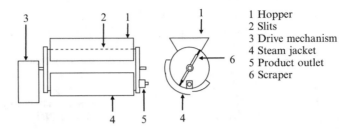

1 Hopper
2 Slits
3 Drive mechanism
4 Steam jacket
5 Product outlet
6 Scraper

Fig. 14.6 *Khoa* making machine

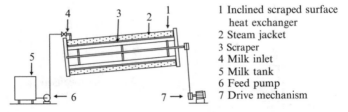

1 Inclined scraped surface
   heat exchanger
2 Steam jacket
3 Scraper
4 Milk inlet
5 Milk tank
6 Feed pump
7 Drive mechanism

Fig. 14.7 Inclined scraped surface heat exchanger (ISSHE)

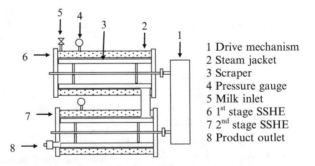

1 Drive mechanism
2 Steam jacket
3 Scraper
4 Pressure gauge
5 Milk inlet
6 1st stage SSHE
7 2nd stage SSHE
8 Product outlet

Fig. 14.8 Two-stage scraped surface heat exchanger (SSHE)

1. *Traditional Method*: Traditionally, *ghee* is produced at household level by fermenting the whole milk into curd and churning the curd to recover fat (indigenous butter) followed by heat clarification at 105–145 °C.
2. *Industrial Process*: In the industrial sector, *ghee* is manufactured by batch processes directly from cream or via cream-butter route or by the prestratification (Punjrath 1991; Mathur 1991).

   (a) *Direct Cream*: A steam jacketed kettle of 500–1000 L capacity is generally employed. Heat clarification is done at 105–120 °C until entire moisture is removed. The curd particles become brown and characteristic *ghee* flavour is developed. After cooling and sedimentation, *ghee* is filtered to remove the sediment before packing in tin containers. Centrifugal *ghee* clarifier can also be used for removal of sediments and residues. Direct cream method yields a higher quantity of ghee residue and takes a much longer time. Cream is at times washed to reduce solids not fat (SNF) also known as *ghee* residue and to improve yield of *ghee*.

(b) *Cream-Butter Method*: This process involves churning of cream to obtain butter and heat clarification is done at 105–110 °C. Rest of the process is same as direct cream method.

(c) *Prestratification*: In the prestratification method, butter is melted at 80 °C and held for 30 min. Three layers are formed. The top layer being scum, middle layer of fat and bottom layer of butter milk serum. Removal of bottom aqueous layer helps in reduction of 70–95 % of the moisture, and also brings economy in steam consumption.

(d) *Continuous Ghee Making Machine*: Abhichandani et al. (1995) designed continuous *ghee* making machine by integrating continuous butter melter with scraped surface heat exchanger (SSHE). Butter (80–85 % fat) at 10 °C was fed into the butter melter and ghee was manufactured at 115–118 °C. Steam consumption was 35 kg per 100 kg of ghee which was 50 % lower compared to 68 kg per 100 kg of ghee required in jacketed steam kettle. Further modification was made and a ghee clarifier/bag filter system was used to remove ghee residues (Abhichandani 1997).

3. *Butter Oil Method*: *Ghee* can be manufactured from butter oil by incorporating curd/skim milk powder. The flavour quality of *ghee* and shelf life are comparable to dairy *ghee*. Curdy flavour can be simulated by employing low temperature clarification (105 °C) and by addition of cultured milk or curd powder at the time of heating (Rajorhia 1995). Cooked flavour can be simulated by clarification at 115 °C for 10 min or 120 °C for 5 min or 125 °C without any holding time.

References

Abhichandani H (1997) Innovation in small scale manufacture of Khoa and Ghee. Processing of International Dairy Federation Workshop on Small Scale Dairy Processing and indigenous milk products, pp 34–43

Abhichandani H, Bector BS, Sarma SC (1995) Continuous ghee making system-design operation and performance. Ind J Dairy Sci 48:646

Agarwal AK (1996) Studies on traditional and centrifugal methods of Paneer production. Ph.D. Thesis. Department of Agricultural and Food Engineering, Indian Institute of Technology, Kharagpur

Agrawala SP, Sawhney IK, Kumar B (1987) Mechanized conical process vat. Indian Patent No. 165440

Aneja RP, Mathur BN, Chandan RC, Banerjee AK (2002) Fat rich products. In: Aneja RP, Mathur BN, Chandan RC, Banerjee AK (eds) Technology of Indian milk products. Dairy India Publication, New Delhi, p 184

Banerjee AK, Verma IS, Bagchi B (1968) Pilot plant for continuous manufacture of Khoa. Ind Dairyman 20(3):81–83

Christie IS, Shah US (1990) Development of a Khoa-making machine. Indian Dairyman 42(5):249–252

DAHDF (2015) Annual report 2013–2014. Department of Animal Husbandry, Dairying & Fisheries (DAHDF). Ministry of Agriculture, Government of India, New Delhi

Das H (2000) Mechanized processing of Indian dairy products: Chhana, Paneer, Sandesh and Rasogolla. Indian Dairyman 52(12):83

Das S, Das H (2009) Performance of an impact type device for continuous production of Paneer. J Food Eng 95:579–587

Mathur BN (1991) Indigenous milk products of India: the related research and technological requirement. Indian Dairyman 42(2):62–71

More GR (1985) Design of Khoa making plant. J Inst Eng 65 Pt. AG 1–2 August

Pal MA (2002) Processed paneer-contemporisation of ancient delight. Indian Dairyman 54(10):78

Pal D (2005) Role of membrane processing in traditional dairy products. National seminar on value added dairy products. NDRI, Karnal

Pal D, Raju P (2007) Process modifications for the manufacture of Indian traditional dairy products from buffalo milk. In: Proceedings of International Conference on Traditional Dairy Foods, 14–17 Nov 2007, pp 132–140

Pal D, Garg FC, Verma BB, Mann M (2002) Application of selected membrane systems for improving the quality of traditional Indian Dairy Products. NDRI Annual Report (2001–2002), pp 39

Patel RK (1991) Indigenous milk products of India. Indian Dairyman 43(3):123

Patil GR (2005) Innovative processes for indigenous dairy products. Indian Dairyman 57(12):82

Punjrath JS (1991) Indigenous milk products of India: The related research and technology requirements in process equipment. Indian Dairyman 42(2):75–86

Punjrath JS, Veeranjanyalu B, Mathunni ML, Samal SK, Aneja RP (1990) Inclined scraped surface heat exchanger for continuous Khoa-making. Indian J Dairy Sci 43(2):225–230

Rajorhia GS (1995) Traditional Indian dairy products technologies: ready for mass adoption. Indian Dairyman 47(12):5

Rajorhia GS, Indu S, Srinivasan MR (1984) Use of ionizing radiation for sterilization of packaging materials for dairy products. Asian J Dairy Res 3(2):91

Rajorhia GS, Pal D, Garg FC, Patel RS (1991) Evaluation of the quality of Khoa prepared from different mechanized systems. Indian J Dairy Sci 44(2):181–187

Sachdeva S, Patel RS, Kanawjia SK, Singh S, Gupta VK (1993) Paneer manufacture employing ultrafiltration. 3rd International Food Convention, IFCON, Mysore

Sachdeva S, Patel RS, Tiwari BD, Singh S (1994) Manufacture of Chakka from milk concentrated by reverse osmosis. 24th International Dairy Congress, Melbourne, Australia, Jb. 36, 415

Sharma DK (1998) Ultrafiltration for manufacture of indigenous milk products: Channa and Shrikhand. Indian Dairyman 50(8):33–37

Sharma DK, Reuter H (1991) A method of Chhana making by ultrafiltration technique. Indian J Dairy Sci 44:89

Sharma DK, Reuter H (1992) Ultrafiltration technique for Shrikhand manufacture. Indian J Dairy Sci 45:209

Sharma N, Agrawala SP, Singh RRB (2003) New processes and equipments for manufacture of Indian milk products and future strategies. Paper presented at the National Workshop on Identification of Technologies and Equipments for Meat and Milk Products organised during 5–6 Sept 2003 at Indian Veterinary Research Institute, Izatnagar, Bareilly

Sindhu JS (1996) Suitability of buffalo milk for products' manufacturing. Indian Dairyman 48(2):41–47

Singh MD (1994) Studies on continuous acid coagulation of buffalo milk. Ph.D. Thesis, Indian Institute of Technology, Kharagpur

Verma RD, Dodeja AK (2000) Development of equipments for the manufacture of indigenous dairy products. Indian Dairyman 52(10):23–24

Chapter 15
Yunnan Pu-erh Tea

Jiashun Gong, Qiuping Wang, Sarote Sirisansaneeyakul, and Anna McElhatton

Contents

J. Gong (✉)
Faculty of Food Science and Technology, Yunnan Agricultural University,
Kunming 650201, P.R. China
e-mail: gong199@163.com

Q. Wang
Faculty of Food Science and Technology, Yunnan Agricultural University,
Kunming 650201, P.R. China

Department of Biotechnology, Faculty of Agro-Industry, Kasetsart University,
50 Ngam Wong Wan Road, Chatuchak, Bangkok 10900, Thailand
e-mail: soffywang87@gmail.com

S. Sirisansaneeyakul
Department of Biotechnology, Faculty of Agro-Industry, Kasetsart University,
50 Ngam Wong Wan Road, Chatuchak, Bangkok 10900, Thailand
e-mail: sarote.s@ku.ac.th

A. McElhatton
Faculty of Health Sciences, University of Malta, Msida, Malta
e-mail: anna.mcelhatton@um.edu.mt

© Springer Science+Business Media New York 2016
A. McElhatton, M.M. El Idrissi (eds.), *Modernization of Traditional Food Processes and Products*, Integrating Food Science and Engineering Knowledge Into the Food Chain 11, DOI 10.1007/978-1-4899-7671-0_15

15.1   Introduction

15.1.1   History of Yunnan Pu-erh Tea in China

During Ming Dynasty (1368 C.E.–1644 C.E.), Kunming Taihua tea, Dali Gantong temple tea and Wanding (Changning country) tea were the most known teas of the period. Pu-erh tea has a much longer history that can be traced back to the Eastern Han Dynasty (25 C.E.–220 C.E.). In those times, tea was processed close to where it was cultivated and shipped directly to markets. There were various varieties such as "scissors crude tea" from Yongning (now known as Ninglang county) which got its name because it was actually harvested using scissors, "Pu-erh tea" from Cheli (presently known by the names of Pu-erh county and Xishuangbanna) and "Wumeng tea" from Wumeng (presently Zhaotong city).

"Pu-erh tea" was the dominant variety and was widely distributed in Yunnan province. "DianhaiYu Hengzhi", a book written in the 4th-year Jiajing (1525 C.E.) of Ming Dynasty, describes the prosperity that the tea trade brought and provides a description of the trade:

> The six Tea Mountains for producing Pu-erh tea are Youle, Gedeng, Yibang, Mangzhi, Manzhuan and Mansa. Hundreds of thousands people work on the Tea Mountains to harvest and process Pu-erh tea. Then the Pu-erh tea is bought and transported to the other places by the tea merchants on the horsebacks. All of the roads are full of caravans, and the tea merchants make a tidy profit from the Pu-erh tea business.

In the 13th year (1748 C.E.) of Qianlong emperor during Qing Dynasty, the Tibetan people consumed large quantities of Yunnan tea, and the Yunnan tea sales to Tibet were so significant that the tea so traded was given its own identity and called Chitsu Pingcha. The Chitsu Pingcha was a unit of tea that consisted of a case of seven circles or cakes, with a weight of 49 Liang (equivalent to approximately 1.83 kg). According to the historical records, the annual sales volume of Chitsu Pingcha was said to reach 3000 picul (approximately 150 t) at that time. By 1825, Ruan Fu, a writer of Qing Dynasty, wrote in his book on Puerh tea that "this tea was well known globally". At the end of the nineteenth century (1894), the Yunnan province tea sales reached 1500 t. At that time, a fine variety of *Camellia sinensis* Kuntze. var. *assamica* Kitamura was initially introduced to many other areas of western Yunnan for plantation. In 1937, new tea-growing regions were extensively developed in Yunnan province. As a result, the entire province produced Pu-erh tea with harvests reaching some 9800 t annually. In 1940, the Yunnan tea sold to Tibet reached its peak of 40,000 packages (about 1.5×10^6 kg).

The history of Yunnan Pu-erh tea, also has ties with the well-known "Tea Horse Road" (Cha Ma Gu Dao) and The "Silk Road" that are both networks of land trading routes. The "Silk Road" connected north western China to Europe from the times of the Han dynasty (25 C.E.–220 C.E.), while the "Tea Horse Road" connected China to various parts of Asia and Europe before seafaring became. It should be noted that the "Tea Horse Road" is sometimes referred to as the "Silk Road of the south".

The "Tea Horse Road" got its name as Chinese tea and horses were the main products traded along the route (together with medicine, salt, cloth, and skins, mostly carried by mules). Historians have traced the origins of the "Tea Horse Road" back to the Tang dynasty (618 C.E.–907 C.E.), when tea was transported out of Yunnan to Beijing, Tibet, and other Southeast Asian countries. The "Tea Horse Road" was further developed during the Song (960 C.E.–1279 C.E.) and Ming dynasty (1368 C.E.–1644 C.E.), and remained a key trading route for Pu-erh tea and other commodities until the Qing dynasty (1644 C.E.–1911 C.E.). There were five routes of the "Tea Horse Road", they were specifically depicted as follows:

Guan Ma route, which ran from Pu'er to Kunming, and further to other provinces within mainland China. It was used for transporting imperial tea to Beijing.
Guan Zang route, which ran from Pu'er to Xiaguan, Lijiang, Shangri-la, and into Tibet; and from Tibet to Nepal and other countries.
Jiang Lai route, which ran from Pu'er to Jiangcheng, into Lai Chau of Vietnam, and onto Tibet and Europe.
Dry Season route, which ran from Pu'er to Simao in Yunnan, and onto the Lancang River, Menglian, and Burma.
Meng La route, which ran from Pu'er to Mengla in Yunnan, and onto northern Laos and Southeast Asia.

At that time, sun-dried green tea leaves were packed and shaped into cake, brick, mushroom, square and bowl shapes to be convenient for transportation. Transportation times lasted weeks if not months, as an example the transportation on the Guan Zang from Pu'er to Tibet, took some 100 days during which time the colour of the tea became darker, and flavour no longer astringent. This occurred because the packed green tea leaves underwent oxidation and fermentation during the trade journey; both processes occurred because of the interaction of the tea with moisture and due to temperature fluctuations. This caused the tea's organoleptic properties to gradually change such that they became highly desirable and therefore also so valued.

In the 1930s due to improved transportation, many new tea roads were opened into Myanmar, Laos and India. Meanwhile packaging and the storage conditions in warehouses improved significantly.

Due to the fact that improved transportation processes significantly reduced the time to reach Tibet had reduced from 100 days to 40 days natural fermentation that required longer process times was incomplete. As a consequence of this processing issue various tea processing factories began to artificially ferment Pu-erh tea. This led to the development of the modern manufacturing technique of Pu-erh tea by solid state fermentation (SSF).

15.1.2 Definition of Yunnan Pu-erh Tea and Its Types

Historically various kinds of tea handled by the Pu-erh local government were called the Pu-erh tea. Pu-erh tea had been originally produced in the Diannan Lanchan River basin, using crude sun-dried green tea prepared from the fresh leaves of *Camellia sinensis* var. *assamica*, through a process of "enzyme(s) inactivation of rolling-sun drying" technology to form raw sun-dried green tea and further to shaped and compressed each kind of compressed tea by heating with steam.

In 1973, the Yunnan tea import–export company first developed a new technique for processing modern Pu-erh tea in the Kunming Tea Factory and Menghai Tea Factory. The Chinese Classic Book of Tea stated that Pu-erh tea was produced from sun-dried green tea that then underwent SSF, and then shaped into tea bricks.

In 2008, Pu-erh tea was declared "a product with geographical indications", and the production areas clearly defined and confined to certain regions in Yunnan province specifically, between parallels 21°10′ and 26°22′ north latitude, 97°31′ and 105°38′ east longitude these designated areas included the fields, towns, cities and counties.

Broadly speaking, the Pu-erh tea products on the market can be categorized into several types according to different standards, as shown in Fig. 15.1. Firstly, Pu-erh tea can be divided into Pu-erh raw tea and Pu-erh ripened tea according to the processing technology and the quality characteristics (Fig. 15.1a). The raw tea, a kind of green tea, is made directly from the sun-dried green tea by further autoclaving and a compression process. Its chemical constituents and quality are very similar to those of the sun-dried green tea. Pu-erh ripen tea is normally made from the sun-dried green tea by microbial post-fermentation at higher temperature (about 50 °C) and higher humidity conditions. In addition, the compressed Pu-erh raw tea can be turned into Pu-erh ripened tea after natural ageing during long-term storage which is generally known as Pu-erh aged tea. Secondly, Pu-erh tea can be divided into loose tea and compressed tea according to shape. After the autoclaving and drying processes, the compressed tea can be made by compressing the loose tea into different moulds to make different shapes (Fig. 15.1). Among the types of Pu-erh compressed tea, cake tea is the most common. Other desired shapes of compressed tea can be made as required, e.g., melon-shaped, tuocha-shaped, mushroom-shaped, brick-shaped and column-shaped (Fig. 15.1b). Currently, the dominant type of Pu-erh tea produced and consumed of is the microbial fermented Pu-erh ripe tea.

Microbial solid state fermentation is known to change the composition of the tea. Tea polyphenols, catechins, thearubigins and theaflavins decreased, while amino acids and the carbohydrates content are somewhat reduced. There is a development of fragrance compounds, many of which have mellow fragrance such as nonanal, linalool, oxidized linalool, 1-ethyl-2-formacyl pyrrole and phenylacetaldehyde are obviously high. In addition, gallic acid, 1,2-dimethoxy-4-methyl benzene, 1,2-dimethoxy-4-ethyl benzene and 1,2,3-trimethoxy-4-ethyl benzene noticeably increase. These characteristic profiles demonstrate that, Pu-erh tea should not belong to the block or brick tea, but should have its own tea variety designation because of the uniqueness of its characteristics.

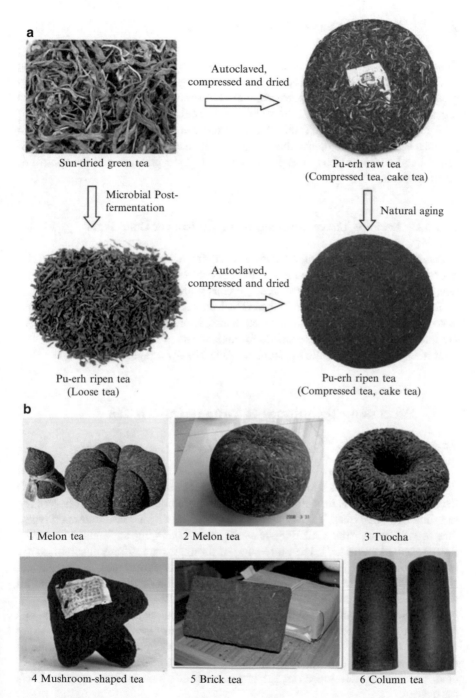

Fig. 15.1 Different types and shapes of Pu-erh tea. ((**a**) Different types and their processing relations; (**b**) various shapes of Pu-erh tea). *Source*: Lv et al. (2013)

15.1.3 Pu-erh Tea as a Functional Food

Zhao Xueming, a writer during the Qing Dynasty, was cited in the Ben Cao Gang
Mu (a Chinese herb pharmacopoeia), as describing Pu-erh tea as "mild and fragrant… it has incomparable perfume, which is very good for drinkers, helping in
digestion and eliminating expectoration" (Chen et al. 2010).

A review by Zhang et al. (2005) has reported that Pu-erh tea has the potential to
produce various physiological changes that include reduction of blood lipids, cholesterol and control of body weight, lowering of blood pressure and the control of
arthrosclerosis.

15.1.3.1 Avoiding Cancer and Improving Anticancer Conditions

In addition studied the potential for Pu-erh tea effects on cancer cells. The tea polyphenols together with their oxidation and degeneration of resulting complex products in Pu-erh tea are all important components with anticancer potential.

In addition, Pu-erh tea prepared from fresh leaves of *Camellia sinensis* var. *assamica* is rich contains, like beta-carrot elements, vitamin B_1, B_2, C, E and so on. All
are important compounds responsible for anticancer.

It is also reported to settle the stomach and to have antiageing effects.

15.2 Processing Techniques of Yunnan Pu-erh Tea

15.2.1 Traditional Techniques for Pu-erh Tea Processing

Although Pu-erh tea has a long cultural history technical information about Pu-erh
tea related literature is limited. It is known that based on the picking or harvesting
time and fresh leaf quality and type, Pu-erh tea can be divided into four main types,
"the wool point", "the tender tea-leaves", "the Xiaomang tea" and "the valley
flower-scented green tea". The processing of these teas are different, subsequently
they also differ. The finished products are described as tight round tea, female tea,
Jin Yuetian and lump tea. A comprehensive description of the predecessors of all
forms of Pu-erh tea processing is as shown in Fig. 15.2.

The Pu-erh tea produced in the traditional way acquires its shape due to the different types of basket used to transport and export it to places such as Tibet, Hong
Kong and Myanmar. This traditional processing method keeps is still in use. Sun-
dried green tea is used as raw materials; hence it is classified as green tea. The tea's
characteristics are developed through steam rubs, in which the temperature affects
tea quality.

Fig. 15.2 The traditional procedure of processing Pu-erh tea

Fresh leaves

(*Camellia sinensis* var. *assamica*)

↓

Fixing

↓

Hand rolling

↓

Sun drying

↓

Sun-dried green tea

(Starting raw materials)

↓

Steaming and pressing

↓

Variety of Pu-erh tea

↓

Slow oxidation

↓

Pu-erh tea

It is well accepted that the quality attributes were traditionally formed during the long transportation times. Therefore, Pu-erh tea quality is undoubtedly the product of the specific historical conditions.

15.2.2 Modern Techniques for Pu-erh Tea Processing

Economic change has motivated the development of the modern technique for the Pu-erh tea production. In the twentieth century, at the beginning of 1970s, the consumers' request for quality Pu-erh tea increased. The Yunnan Province tea companies began to ferment and reprocess Pu-erh tea such as had been done in the

Kunming tea factory and Menhai tea factory. Pu-erh tea processing entered the new development era. The whole process of studying Pu-erh tea, from raw materials, processing methods, chemical composition, quality testing, were developed and the scientific idea of modern Pu-erh tea was finally established.

The modern Pu-erh tea, using sun-dried green tea as raw materials, through splashing water, solid state fermentation, mellow stage and drying and so on, made Pu-erh tea a special kind of tea. The traditional characteristics of Pu-erh tea namely its mellow taste and Chen fragrant characteristic were retained in modern processing.

Pu-erh tea became widely available especially in Hong-Kong, Macao and Taiwan, and further afield in Southeast Asia and the volume of exports continue to increase. The microbial involvement in the process is complex and is the source of the characteristic tea quality that has been associated with its cultural history. Figure 15.3 describes the modern technical process of manufacturing Pu-erh tea. In this modern processing method, there are four stages that have evolved from the traditional craft link:

1. Pu-erh tea's raw materials, which are typically prepared from *Camellia sinensis* var. *assamica* of Yunnan Province. After enzymes inactivation of fresh tea leaves by hot steam, the steps of rolling tea leaves and drying them by sunlight are essential to obtain the sun-dried green tea.
2. Solid state fermentation (SSF), the SSF facilitates the slow oxidation of Pu-erh tea. The fermentation involving beneficial microorganisms which through their growth and metabolism induce characteristic biotransformation, forming the unique quality characteristics of Pu-erh tea.
3. Special climatic conditions, in Yunnan Province, naturally provide the optimal conditions for the requisite microbial secondary metabolism that facilitates the production of quality Yunnan Pu-erh tea. The sunshine in Yunnan appropriate for the production of Pu-erh tea, where the yearly average temperature ranges from 12 to 23 °C providing the solar energy to sun-dry green tea leaves.
4. Pressing and shaping, the hot and damp material is also thought to influence the quality of Pu-erh tea. The pressing and shaping may influence chemical transformation of ingredients to remove astringency and render the tea mellow in taste.
5. Storage and ageing, under controlled evidence based storage conditions, indicate that the longer the Pu-erh tea is aged the mellower in taste it becomes. The long storage time may cause changes in the chemical composition of Pu-erh tea through oxidation. Due to the fact that there is significant variability in the raw materials, the finished Pu-erh tea typically also varies and are given names such as the raw cake, "shengpu series" and the modern craft of the ripe cake, "shupu series", as well as "dry store Pu-erh tea" and "wet store Pu-erh tea" that are formed by different storage conditions. As a result, there are many kinds of Pu-erh tea.

Large fresh leaves in Yunnan Province

(*Camellia sinensis* var. *assamica*)

↓

Fixing

↓

Rolling

↓

Sun drying

↓

Sun-dried green tea

↓

Water spraying

↓

Inoculation with microorganisms

↓

Solid state Fermentation (SSF)

↓

(i) Unpacked Pu-erh tea, and/or (ii) Shaped Pu-erh tea (by pressing and shaping)

↓

Packaging

↓

Storage and aging

↓

Yunnan Pu-erh tea

Fig. 15.3 Modern procedure for producing Pu-erh tea

15.3   Research Progress in Chemical Composition
of Yunnan Pu-erh Tea

The chemical composition of tea is the foundation of tea quality. So far, there are about 600 known compounds in tea of which more than 450 are organic compounds kinds. With the exception of carbohydrate, fat and lipid, found in the raw materials, the other compounds are all the secondary metabolites. Among tea's secondary metabolites, there are polyphenols and purines that account for 20–38 % and 3–5 % of the tea's composition, respectively. In general, tea colour, its smell taste, and the quantity and profile of the secondary metabolites are all factors affecting the decision regarding quality. The tea infusion quality is therefore an association of various attributes that include the taste of the tea infusion that is influenced by chemical components and the processing itself which gives the tea its full-bodied taste, its colouration (as shown in Fig. 15.4), and its Chen (aged) or mellow fragrance.

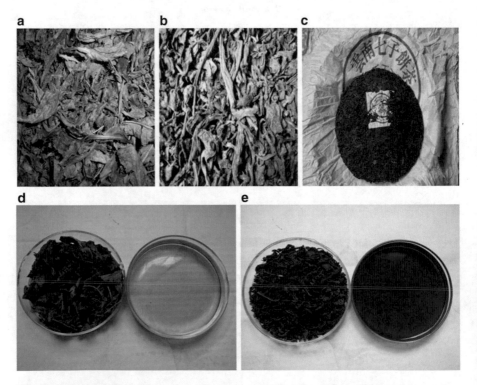

Fig. 15.4 Comparison between sun-dried green tea and Pu-erh tea after solid state fermentation, (**a**) sun-dried green tea, (**b**) Pu-erh tea after solid state fermentation, (**c**) Yunnan Chitsu Pingcha, as a kind of Pu-erh tea, (**d**) a water extract colour from sun-dried green tea and (**e**) a water extract colour from Pu-erh tea after solid state fermentation

15.3.1 Chemistry of Raw Materials

Pu-erh tea is prepared from sun-dried green tea as starting raw materials. After sun-drying the fresh leaves of *Camellia sinensis* var. *assamica*, a step of special solid state fermentation is required, then pressing and shaping is particular to produce a variety of Pu-erh tea. For sun-dried green tea, its production is mainly from the area of Yunnan national minority. The fresh leaves of *Camellia sinensis* var. *assamica* are collected by labour picking, inactivation of enzymes, rolled and cooled, and then dried with solar energy.

The moisture content of sun-dried tea leaves is approximately 10–12 %, which is much higher than that of fried green tea (approximately 6–8 %). The modern Pu-erh tea is basically evolved from the traditional Pu-erh tea. The typical characteristics are mellow taste and remarkable Chen fragrant. However, both are commonly so-called Pu-erh tea, neither traditional nor artificial Pu-erh tea is of necessity identified. With using big leaf planting tea as material and fermenting step result in the characteristic quality of brown colour and Chen fragrant. Since the natural fermentation is quite slow process for traditional Pu-erh tea, the fermentation time of modern Pu-erh tea is fast under artificially controlled conditions. Nevertheless, both the techniques form similarity of Pu-erh tea characteristic quality.

Pu-erh tea is well-known worldwide at the present time. The production and exporting volume of Pu-erh tea are increasing of great concern. The unique quality characteristics of Pu-erh tea are essentially formed, because particular chemical substances present in bud leaf and tender stem are typical, additionally the biochemical changes after the solid state fermentation process are of great importance.

Table 15.1 summarizes chemical components of various Chinese sun-dried green teas (Gong et al. 2005). They are especially rich in polyphenols and a number of different oxidized products depending upon the source of the leaves. The raw materials from platform (mesa) plantation area tea show higher contents of polyphenol, total catechins, oligosaccharides and total carbohydrates than those from old tea, except those of flavanones, where the theaflavins and thearubigins are actually lower. This has laid the foundation for the unique quality formation of Pu-erh tea. Zhou and Yang (2000)) reported that 20 phenolic compounds together with caffeine were isolated from crude materials of Pu-erh tea produced in Yunnan Province, China. They were identified by spectroscopic method as follows: (–)epicatechin, (–)epigallocatechin, (–)epigallocatechin-3-O-gallate, (–)epicatechin-3-O-gallate, (–)epiafzelechin-3-O-gallate, (+)catechin, (±)gallocatechin, quercetin, quercetin-3-O-β-D-glucopyranoside, rutin, kaempferol, kaempferol-3-O-β-D-glucopyranoside, kaempferol-3-O-rutinoside, strictinin, 1,6-O-digalloyl-β-D-glucopyranose, theogallin, chlorogenic acid, 3-α,5-α-dihydroxy-4-α-cafeoyl-quinic acid, coniferin, gallic acid. Their chemical structures are shown in Fig. 15.5.

Table 15.1 Chemical components of various Chinese sun-dried green teas

Sun-dried green tea	Chemical components (%)										
	TP	TC	FL	TF	TR	TB	TS	TPS	TOS	TE	Ash
Mangjing	35.11	12.03	1.69	0.14	8.68	2.09	11.00	0.11	7.02	39.60	5.05
Nannuo old tea	37.23	11.26	1.26	0.17	6.93	2.20	10.30	0.36	8.55	38.40	4.75
Manggeng old tea	37.80	12.58	1.60	0.14	6.54	2.11	9.50	0.35	6.55	40.00	5.05
Manggeng old tea	36.05	13.09	1.61	0.15	6.88	1.75	10.42	0.90	7.77	40.40	4.75
Manggeng mesa tea	30.97	11.07	1.84	0.23	9.83	2.66	9.26	0.10	6.50	40.20	5.20
Bangwei old tea	29.90	10.21	1.84	0.18	7.61	2.34	9.32	0.15	8.70	37.60	5.25
Bangwei mesa tea	30.62	10.52	1.83	0.21	8.11	2.37	9.21	0.27	5.94	38.60	5.30
Jingmai old tea	35.98	12.80	1.55	0.13	7.92	1.87	9.48	0.21	7.12	39.00	4.75
Jingmai mesa tea	32.36	11.58	1.75	0.18	9.37	3.06	9.22	0.19	8.66	40.00	5.10

Note: *TP* tea polyphenols, *TC* total catechins, *FL* flavone, *TF* theaflavins, *TR* thearubigins, *TB* theabrownins, *TS* total sugar, *TPS* tea polysaccharide, *TOS* tea oligosaccharide, *TE* tea extracts
Source: Gong et al. (2005)

15.3.2 The Changes of Chemical Components of Pu-erh Tea during Solid State Fermentation

15.3.2.1 The Polyphenols

The polyphenols are important active components in tea leaf and account for the 60–80 %, of active ingredients in tea that are responsible for the expression of tea colour, taste and fragrance. The fermentation process of Pu-erh tea with *Aspergillus niger* changes the chemical profile polyphenols (Table 15.2), caffeine, flavone, thearubigins and soluble oligosaccharides decreased during SSF. The theabrownins and soluble tea polysaccharides increase significantly. While the teaflavins, total soluble carbohydrates and ash were not remarkably changed (Tables 15.2 and 15.4). It is thought that these variations during the fermentation process could be considered as being quality characteristics of Pu-erh tea.

Guo et al. (2001) also reported that the catechins ester decreased markedly more than the other catechins. While, gallic acid also increased significantly (Table 15.3) as did the insoluble tea polyphenols by up to 70–80 % which contribute to the reduction of the bitterness of the tea infusion.

Pu-erh tea infusion characteristically has a red to red-brown colour that persists as a consequence of its chemical profile. The theaflavins are responsible for the bright infusion colour while the thearubigins are associated with both colour and

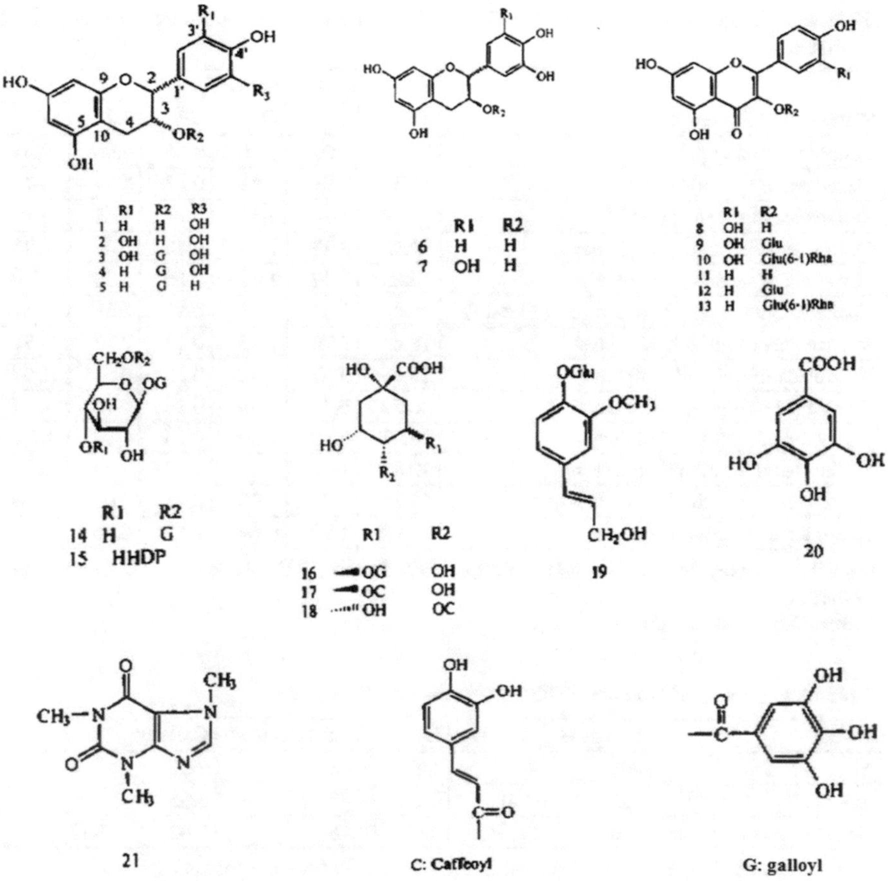

Fig. 15.5 The chemical structures of compounds 1–21. *Source*: Modified from Zhou and Yang (2000)

taste intensity of the infusion. Theabrownins is associated with the darkening of the infusion; when at levels of 6–8 %, the infusion colour appears a bright red-brown, at lower levels (5 %), the infusion colour tends to a bright red-orange colour associated with insufficient fermentation.

In the SSF process, theabrownins levels remain stable while in the other the main chemical constituents of tea are significantly altered (Table 15.2). Moreover, this theabrownins content is considerably high, and considered to be a significant quality attribute of the tea.

Luo et al. (1998) thought that the interactions between polyphenols and production of insoluble protein-polyphenols complex persist and increase as in the tea leaf residues as SSF progresses. Hot and damp conditions that persist inside batches of fermenting tea facilitate the oxidation of tea polyphenols and formation of complex protein-polyphenols complex.

Table 15.2 Chemical components of different upturned tea in the solid state fermentation of Pu-erh tea

Samples and fermentation time	Chemical components (%)					
	TP	TC	FL	TF	TR	TB
Central tea of first pile (10 days)	31.05	9.07	1.46	0.16	4.00	2.35
Mixed tea of first pile (10 days)	33.3	9.84	1.60	0.17	6.41	2.25
Surface tea of second pile (20 days)	24.19	8.12	1.39	0.17	4.61	3.75
Central tea of second pile (20 days)	30.13	7.23	1.58	0.15	6.06	3.23
Lower tea of second pile (20 days)	26.85	7.44	1.42	0.16	5.30	4.14
Cankered bottom tea of second pile (20 days)	26.45	6.64	1.56	0.16	4.8	4.37
Surface tea of third pile (30 days)	18.6	7.73	1.13	0.13	2.34	5.71
Central tea of third pile (30 days)	26.91	5.91	1.42	0.15	4.54	1.93
Bottom tea of third pile (30 days)	26.52	6.4	1.36	0.16	4.76	4.14
Mixed tea of third pile (30 days)	26.52	5.88	1.36	0.16	4.97	4.74
Surface tea of fourth pile (40 days)	12.45	1.97	0.59	0.10	0.28	13.74
Central tea of fourth pile (40 days)	11.18	1.04	0.79	0.11	0.01	11.63
Lower tea of fourth pile (40 days)	13.92	1.90	0.64	0.12	0.26	12.45

Note: *TP* tea polyphenols, *TC* total catechins, *FL* flavone, *TF* theaflavins, *TR* thearubigins, *TB* theabrownins
Source: Gong et al. (2005)

Table 15.3 Content of catechins in different Pu-erh tea (*Unit*: %)

Pu-erh tea	L-EGCG	L-ECG	L-EC	Total catechins	Caffeine	Gallic acid
Yunnan Tuocha	4.65	4.37	trace	9.48	3.82	0.34
Yunnan Pu-erh tea	0.03	0.01	0.21	0.27	3.93	1.72
Yunnan Chitsu Pingcha	0.14	0.19	0.40	0.78	2.82	1.12

Note: *EC* epicatechin, *EGC* epigallocatechin, *EGCG* epigallocatechin gallate
Source: Guo et al. (2001)

The irreversible protein-polyphenols complex is therefore the main reason for the increase of insoluble tea polyphenol content during the SSF process. This is generally caused by quinine (formed from the oxidation of catechins) and the insoluble thearubigins and theabrownins combining with protein. The catechins too, under certain conditions, can also interact with proteins to produce the insoluble complex.

These insoluble protein complexes are reported to be responsible for the reduction of tea extracts from 35.6 to 27 % (Fig. 15.6).

In addition, peroxidase may also promote the oxidation of theaflavins to form insoluble protein complexes. Because the fermentation temperature is quite high approximately 40–60 °C, peroxidase activity reaches optimal levels; as a result, the quantity of theaflavins diminish while that of insoluble thearubigins increase. In conclusion, the production of insoluble protein-polyphenols complex improves the characteristic taste of Pu-erh tea.

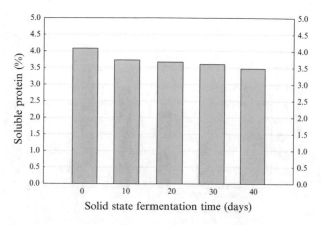

Fig. 15.6 Changing of soluble protein of sun-dried green tea during SSF fermentation. *Source*: Gong et al. (2005)

Table 15.4 Carbohydrates content of different upturned tea during solid state fermentation of Pu-erh tea

Tea samples during SSF	Components (%)				
	TS	TPS	TOS	TE	Ash
Central tea of first pile (10 days)	8.88	0.60	6.51	35.60	5.53
Mixed tea of first pile (10 days)	9.12	0.37	6.51	34.60	5.60
Surface tea of second pile (20 days)	8.31	0.94	6.43	35.53	5.99
Central tea of second pile (20 days)	8.79	1.27	5.49	34.39	5.44
Lower tea of second pile (20 days)	8.39	1.28	4.87	31.40	5.34
Cankered bottom tea of second pile (20 days)	8.79	1.49	4.53	36.67	5.64
Surface tea of third pile (30 days)	9.19	2.25	4.57	33.40	5.53
Central tea of third pile (30 days)	10.02	2.43	5.18	32.89	5.30
Bottom tea of third pile (30 days)	8.72	1.37	4.42	31.82	5.22
Mixed tea of third pile (30 days)	10.24	2.14	4.27	34.93	5.39
Surface tea of fourth pile (40 days)	8.62	3.19	2.18	26.00	5.90
Central tea of fourth pile (40 days)	7.78	3.09	1.66	24.53	5.70
Lower tea of fourth pile (40 days)	8.48	3.79	2.97	27.00	6.12

Note: *TS* total sugar, *TPS* tea polysaccharide, *TOS* tea oligosaccharide, *TE* tea extract
Source: Gong et al. (2005)

15.3.2.2   The Carbohydrates

The tea contains approximately 10–20 % carbohydrates with a profile that is considered to be related to quality. Table 15.4 shows that unfermented Pu-erh tea contains between 9 and 11 % soluble carbohydrates, when fermented by SSF the soluble polysaccharides and oligosaccharides content change.

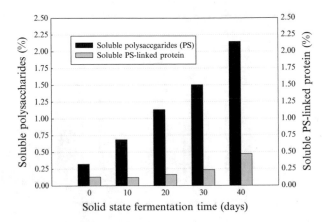

Fig. 15.7 Changing of soluble polysaccharides and polysaccharides-linked protein of sun-dried green tea during SSF fermentation. *Source*: Gong et al. (2005)

The soluble polysaccharides levels change significantly during SSF with the polysaccharide content increasing more than fivefold during SSF (Gong et al. 2005). Moreover, the crude extracts of polysaccharides also contain partially soluble protein (Fig. 15.7) also increase which is consistent with the premise that there is a connection between tea polysaccharides and protein levels.

Luo et al. (1998) indicated the change of soluble carbohydrates content of Pu-erh tea varied during SSF fermentation, after initially falling in concentration the general tendency is for an overall increase in soluble carbohydrate concentration. It has been reported that carbohydrates content decreased because of tea fermentation and storage in a storehouse. For the Yunnan green tea, the content of soluble sugars is 71.10 %, but the soluble sugars in loose Pu-erh tea is 20.3 %, and approximately 36.1 % in Pu-erh Tuocha. The content of soluble sugars being attributed to the growth of microorganisms as soluble sugars can partially serve as the good carbon source for microorganisms' growth.

15.3.2.3   The Changes of Nitrogen Compounds

At the beginning of processing Pu-erh tea, the taste quality is formed by the oxidative degradation of the precursors of the material that will ultimately develop the taste. Caffeine, theabrownins and theine do not vary significantly. Other amino acids such as threonine, glutamic acid and aspartic acid found in fresh leaf tea decreased gradually during SSF, but others such as lysine, phenylalanine, methionine and isoleucine increased during solid state fermentation.

The amino acid profile in the teas are the key factors that determine quality and taste of the infusion. The combination of amino acids and polyphenols provide the characteristics of the tea. Japanese scientists have investigated the composition and contents of amino acids in Pu-erh tea and have shown that every 100 g of Pu-erh tea contained some 16 mg of amino acids. The hydrolysis of protein increased the content of soluble amino acids at the beginning of fermentation time; however, due to microbial growth facilitated by long periods of processing in hot and damp environments, total amino acids decreased. The chemistry involved is thought to involve the partial decomposition of amino acids, oxidation, hydrolysis and deaminizing, and use by microorganisms as nitrogen sources. The SSF process therefore causes shifts in the amino acid profile.

15.3.2.4   Variation of Aromatic Substances

The formation of the characteristic fragrance is especially complicated. It is known that the content of aromatic substances in fresh leaf tea is 0.03–0.05 % that does change during the manufacturing process of Pu-erh tea.

During SSF of Pu-erh tea, many factors such as enzyme catalysis, oxidation and reduction of catechins, moisture and acidic conditions all have the potential to create aromas. From studying Chinese brick tea, Kawakami (1987) identified 132 kinds of fragrance ingredients from brick tea samples. It was thought that this process could be manipulated by controlling microbial fermentation and oxidation to produce desirable aromas.

Chen fragrance components in different raw materials of Pu-erh tea are totally different (Gong et al. 2005). According to Liu and Ina (1987), the main Chen fragrance of Yunnan Pu-erh tea were mainly caused by oxidized linalool (I), oxidized linalool (II), oxidized linalool (IV), linalool, 1,2-dimethoxy-4-ethyl benzene, 1,2-dimethoxy-4-methyl benzene, 1-ethyl-2-formacyl pyrrole, α-terpineol, benzaldehyde, trans-2-trans-4-heptadienal, n-nonanal, n-decanal and other unknown compounds. It is also indicated that the increased contents of octadiene alkone, heptadiene alkone and pentenol were the components that produced Chen fragrance. In addition, the increased fragrant components of 1,2-dimethoxy-4-methyl benzene, 1,2-dimethoxy-4-ethyl benzene and 1,2,3-trimethoxy-4-ethyl benzene of Pu-erh tea resulted from the hydroxylation of gallic acid and the methylation of acyl gallnut.

15.3.2.5   Variation of Tea Extracts

Tea extract is obtained when tea leaves are soaked in hot water; it is the major source of taste and aroma in tea infusions. The high content of tea extract reflects high soluble matter containing tea that to a certain extent is a determinant of the tea quality.

Traditionally processed Pu-erh tea has its unique style. Pu-erh tea is classified as a post-fermented tea and contains soluble carbohydrates and pectin and hydrolysed products that are formed during the fermentation process, which enhance the taste of tea infusion. Shao et al. (1994) reported that the content of tea extracts is very much related to the quality of Yunnan Pu-erh tea. As shown in Table 15.5, the content of tea extracts of Pu-erh tea depends on the solid state fermentation time.

Table 15.5 Variation of tea
extracts during solid state
fermentation of Pu-erh tea

Tea samples during SSF fermentation	Tea extracts (%)
Central tea of first pile (10 days)	35.60
Mixed tea of first pile (10 days)	34.60
Surface tea of second pile (20 days)	35.53
Central tea of second pile (20 days)	34.39
Lower tea of second pile (20 days)	31.40
Cankered bottom tea of second pile (20 days)	36.67
Surface tea of third pile (30 days)	33.40
Central tea of third pile (30 days)	32.89
Bottom tea of third pile (30 days)	31.82
Mixed tea of third pile (30 days)	34.93
Surface tea of fourth pile (40 days)	26.00
Central tea of fourth pile (40 days)	24.53
Lower tea of fourth pile (40 days)	27.00

Source: Gong et al. (2005)

The longer fermentation time shows the less content of tea extracts of Pu-erh tea. Therefore, during solid state fermentation, the fermentation time is practically a key factor controlling the tea extracts.

15.3.3 The Changes of Chemical Composition of Pu-erh Tea during Storage

Through the analysis of Pu-erh tea that had different storing times and which had been sourced from different producer sources, it was possible to establish that storing time does not play particularly for the physicochemical changes of tea components (Table 15.6). It is clear that the changing components of Pu-erh tea have no significant correlation with the longer storing time; therefore, any changes may be caused by differences in processing methods, raw materials used and storage conditions (Gong et al. 2005). It is therefore quite difficult to determine the age of Pu-erh tea through the changes of chemical components. The means to characterize the age of Pu-erh tea based on scientific means is therefore a challenge.

15.4 Profiles of Microorganisms on the Quality of Pu-erh Tea

Chen et al. (1985) and Hu (1979) independently identified a number of microorganisms, namely *Aspergillus niger*, *Penicillium*, *Rhizopus*, *Aspergillus glaucus*, *Saccharomyces* and *Cladosporium* involved in the processing Pu-erh tea. Among them, *Aspergillus niger* was found in approximately 80 % of the sources tested.

Table 15.6 Chemical components of crusted Pu-erh tea

Yunnan Pu-erh tea samples	Content of component (%)												
	TP	TC	FL	TF	TR	TB	TS	TPS	TOS	TE	Ash	TP
Pu-erh tea (8-class) (1984)	14.34	1.98	3.77	0.27	4.52	9.21	11.15	3.39	3.49	30.51	5.10	14.34
Pu-erh tea 79072 (1989)	12.92	2.04	3.31	0.23	3.44	8.27	9.51	2.64	2.59	24.19	5.80	12.92
Chitsu Pingcha (1997)	12.11	1.35	2.49	0.17	1.98	9.75	8.65	2.63	2.31	36.62	7.16	12.11
Brick tea (1998)	8.31	1.11	1.79	0.18	1.21	11.42	7.74	nil	2.50	31.15	7.58	8.31
Chitsu Pingcha (1999)	12.34	1.48	3.52	0.16	2.42	8.93	8.67	2.61	2.29	36.66	7.09	12.34
Tuocha (2000)	10.62	1.10	1.98	0.15	2.24	10.85	8.21	3.23	1.36	34.50	6.87	10.62
Tuocha (2002)	12.25	1.35	2.52	0.19	2.81	9.53	9.36	3.13	1.49	37.15	6.98	12.25
Pu-erh tea (2003)	12.51	1.10	2.04	0.16	2.24	9.81	9.33	2.38	2.74	32.84	7.10	12.51
Pu-erh tea (2-class) (2003)	10.36	1.07	1.58	0.34	2.24	9.90	8.80	2.59	1.98	22.00	6.25	10.36
Pu-erh tea (4-class) (2003)	10.74	1.51	1.4	0.262	1.71	12.76	9.08	2.77	2.48	25.20	6.40	10.74
Pu-erh tea (7-class) (2003)	13.76	2.21	2.39	0.29	3.36	10.56	10.16	3.39	3.11	27.87	5.40	13.76
Pu-erh tea (8-class) (2003)	16.14	2.81	2.22	0.32	4.55	9.13	11.32	3.47	3.91	30.13	5.45	16.14
Old Tuocha (2004)	15.54	3.52	3.87	0.16	1.37	12.25	10.78	0.31	3.26	35.20	5.95	15.54

Note: *TP* tea polyphenols, *TC* total catechins, *FL* flavone, *TF* theaflavins, *TR* thearubigins, *TB* theabrownins, *TS* total sugar, *TPS* tea polysaccharide, *TOS* tea oligosaccharide, *TE* tea extract

Source: Gong et al. (2005)

Observations of SSF revealed that in the early phases of SSF rapid growth and propagation of mesophilic moulds such as *Aspergillus niger* was observed. However, later the SSF process, under dry environments psychrophiles such as *Aspergillus glaucus* started to grow. At the same time, various enzymes underwent reactivation that then caused further chemical changes in the tea composition profiles (Hu 1957, 1979).

Other recent studies by Zhou and Zhao on microbes during the solid state fermentation process of Yunnan Pu-erh tea showed that in addition to the Microorganisms identified independently by Hu and Chen other Aspergilli namely *Aspergillus terreus*, *Aspergillus candidus* and other bacteria were also significant microbes of Pu-erh tea. By far *Aspergillus niger* was the largest contributor. Another significant microorganism identified was *Saccharomyces* sp. These microbes all play direct and indirect roles more or less on quality formation of Pu-erh tea (Zhou et al. 2004; Zhao and Zhuo 2005).

15.4.1   Aspergillus niger

Aspergillus niger is a generally significant industrial microorganism and specifically has been identified as a key organism in the production of Pu-erh tea. Its growth is beneficial to the quality improvement of Pu-erh tea. *Aspergillus niger* belongs to *Deuteromycotina, Hyphomycetes, Moniliales, Moniliaceae* and *Aspergillus*. During solid state fermentation of Pu-erh tea, *Aspergillus niger* may secrete some twenty (20) types of hydrolytic enzymes (Zhao and Zhou 2003). Among them, glucoamylase, cellulases and pectinase decompose many insoluble organic compounds such as polysaccharides, fat, protein, natural fibre and pectin.

After hydrolysis of macromolecular compounds, the micromolecular compounds produced are monosaccharides, amino acids, hydrated pectin and soluble carbohydrates. The organic acids produced by enzymatic activity may promote the development of the desired taste such as the mellow quality characteristics of tea infusion.

15.4.2   Penicillium

Penicillium belongs to *Deuteromycotina, Hyphomycetes, Moniliales* and *Moniliaceae*. *Penicillium,* isolated from Pu-erh tea, can produce a number of enzymes such as glucose oxidase and secrete organic acids including gluconic acid, citric acid and antiscorbutic acid. It also may produce penicillin to eliminate and suppress the growth of putrefactive bacteria; therefore, *Penicillium* is considered to improve the tea quality.

15.4.3 Rhizopus

Rhizopus belongs to *Rhizopus, Zygomycotina, Zygomycetes, Mucorales* and *Mucoraceae*. It can be used to ferment Pu-erh tea (Liu and Ina 1987). It can also produce a number of enzymes such as amylase, sugar forming enzymes, pectinase, proteinase and zymase. These enzymes are active at temperatures ranging from 32 to 40 °C and pH 4.5–6. Among them, the activity of amylase is higher than those of other enzymes. Amylase secreted by *Rhizopus* can catalyse starch in the Pu-erh tea and produce organic acids such as fumaric acid, lactic acid and succinic acid. *Rhizopus* can also produce fragrant ester matters, by transforming sterol race compounds to their derivatives. So, *Rhizopus* is an important organism for Pu-erh tea. In addition, because of its pectinase activity, *Rhizopus* can break down tea leaves during solid state fermentation. Therefore, during solid state fermentation, by controlling suitable temperature and humidity for the growth of *Rhizopus*, this will be beneficial to form the sweet and mellow taste of Pu-erh tea.

15.4.4 Saccharomyces *Yeast*

The *Saccharomyces* yeast belongs to the *Ascomycetes, Hemiascomycetes, Endonycetales* and *Saccharomyces*. It is an important strain which is also responsible for forming the quality of Pu-erh tea. In the fermentation process of Pu-erh tea, the function of hot and damp provides suitable environment for *Saccharomyces* growth and its metabolic activity. This also strengthens the activities of enzymes which result in the changes of tea composition.

In addition, many of the monosaccharides and soluble oligosaccharides produced by *Aspergillus niger* also can be supplied as carbon source for the growth of this yeast. It was indicated that provision of suitable conditions for rapid *Saccharomyces* growth improved the quality of Pu-erh tea with sweet and mellow characteristics (Gong et al. 2005).

Interestingly, Pu-erh tea with the quality characteristic of Chen fragrance (mellow and sweet) is related to the rapid growth of *Saccharomyces* in solid state fermentation of Pu-erh tea. Therefore, appropriate control in the quantity of *Saccharomyces* in processing Pu-erh tea may result in the increasing of those effective nutrients and healthy components of Pu-erh tea, which provide the unique qualities and style of Pu-erh tea. On the other hand, if such control is not appropriate, there could be deterioration of taste and therefore quality of Pu-erh tea.

15.4.5 Bacteria

In a mixed matrix, bacterial growth dominates in conditions of higher temperature and humidity together with a rich source of organic material. Such conditions will produce off-odours and or a rancid taste. However, during the processing of Pu-erh tea, bacterial populations are particularly low and so far the presence of pathogenic bacteria has not been reported. This scenario might be due to growth competition among microorganisms and repression of bacteria by the tea polyphenols.

15.5   Research on Safety of Pu-erh Tea

Pu-erh tea, as a traditional beverage of China, has a long history of culture. They have been drinking Pu-erh tea without particular known risk. However, in a view of the fact that the processing of Pu-erh tea deals with a number of microorganisms a significant number of microbial metabolites and complex compounds are produced. The potential for health hazards is not entirely unfounded; therefore, the chemical toxicity of Pu-erh tea has been evaluated and reports show that in brick format there is ample proof of safety (Liu et al. 1996; Chou et al. 1999; Cao et al. 1998, 2001, 2003; Wong et al. 2003).

Liu et al. (2003), at Southwest Agricultural University, studied on acute toxicity of three typical Pu-erh teas and one baked green tea from Yunnan Province and concluded that all four kinds of tea are highly safe for drinking.

Sun and Liu (2002), from Taiwan University reported that *Aspergillus* aflatoxin was not separated from Pu-erh tea inoculated with aflatoxin-producing *Aspergillus* during simulated production; it seemed that aflatoxin-producing *Aspergillus* could not survive under microflora system of Pu-erh tea fermentation, where *Aspergillus niger* and *Saccharomyces*, the dominant microorganisms, suppressed the growth of harmful microorganisms.

References

Cao J, Zhao Y, Liu JW (1998) Safety evaluation and fluorine concentration of Pu'er brick tea and Bianxiao brick tea. Food Chem Toxicol 36(12):1061–1063

Cao J, Zhao Y, Liu JW (2001) Processing procedures of brick tea and their influence on fluorine content. Food Chem Toxicol 39(9):959–962

Cao J, Zhao Y, Liu JW, Xirao R, Danzeng S, Daji D, Yan Y (2003) Brick tea fluoride as a main source of adult fluorosis. Food Chem Toxicol 41(4):535–542

Chen ZD, Liu QJ, Zhou CQ (1985) The microorganism and the Pu-erh tea ferment. J Tea Sci Technol 4:4–7, In Chinese

Chen YS, Liu BL, Chang YN (2010) Bioactivities and sensory evaluation of Pu-erh teas made from three tea leaves in an improved pile fermentation process. J Biosci Bioeng 109(6):557–563

Chou CC, Lin LL, Chung KT (1999) Antimicrobial activity of tea as affected by the degree of fermentation and manufacturing season. Int J Food Microbiol 48(2):125–130

Gong JS, Zhou HJ, Zhang XF, Song S, An WJ (2005) Changes of chemical components in Pu'er tea produced by solid state fermentation of sundried green tea. J Tea Sci 25(3):126–132

Guo WF, Wu D, Sakata K, Luo SJ, Yang XF (2001) Proceedings of 2001 international conference on O-CHA [tea] culture and science (Session II), 5–8 Oct, Shizuoka, Japan, pp 288–291

Hu JC (1957) Separation and identification of microorganism from four kinds of Bianxiao tea. J Tea 2:20–22, In Chinese

Hu JC (1979) Study on the *Aspergillus* of Pu-erh tea. J Tea 2:20–22, In Chinese

Kawakami B (1987) Aroma characterization for pile tea, Chinese brick tea and black tea. Nihonnokeikagakukaishi 61(4):457–465, In Japanese

Liu QJ, Ina K (1987) Study on the aroma components of Pu-erh tea. J Southwest Agric Univ 9(4):470–475, In Chinese

Liu CH, Hsu WH, Lee FL, Liao CC (1996) The isolation and identification of microbes from a fermented tea beverage, Haipao, and their interactions during Haipao fermentation. Food Microbiol 13(6):407–415

Liu QJ, Chen WP, Bai WX, Li QZ (2003) Acute toxicity evaluation of Pu-erh tea. J Tea Sci 23(2):141–145, In Chinese

Luo LX, Wu XC, Deng YL, Fu SW (1998) Variations of main biochemical components and their relations to quality formation during pile-fermentation process of Yunnan Pu-erh tea. J Tea Sci 18(1):53–60, In Chinese

Lv HP, Zhang YJ, Lin Z, Liang YR (2013) Processing and chemical constituents of Pu-erh tea: a review. Food Res Int 53(2):608–618

Shao WF, Cai X, Yang SR, Yuan W (1994) A preliminary study on the relation between the contents of chemical constituents and quality of Yunnan Pu'er tea. J Yunnan Agric Univ 9(1):17–22, In Chinese

Sun LX, Liu HQ (2002) Anti- arteriosclerosis function of Pu-erh tea. In: 2002 Year China Pu-erh Tea International Conference. Yunnan People's Publishing Press, pp 309–317

Wong MH, Fung KF, Carr HP (2003) Aluminium and fluoride contents of tea, with emphasis on brick tea and their health implications. Toxicol Lett 137(1–2):111–120

Zhang DY, Shi ZP, Liu YL (2005) Development of pharmacological function of Pu-erh tea. Fujian Chaye 1:12–13, In Chinese

Zhao LF, Zhou HJ (2003) Preliminary study on function of yeast in developing Pu-erh tea quality. Tea Garden 2:4–6, In Chinese

Zhao LF, Zhuo HJ (2005) Study on the main microbes of Yunnan Pu-erh tea during pile-fermentation process. J Shandqiu Teach Coll 21(2):129–133, In Chinese

Zhou ZH, Yang CR (2000) Chemical constituents of crude green tea, the material of Pu-erh tea in Yunnan. Acta Botanica Yunnanica 22(3):343–350, In Chinese

Zhou HJ, Li JH, Zhao LF, Hart J, Yang XJ (2004) Study on main microbes on quality formation of Yunnan Pu-erh tea during Pile-fermentation process. J Tea Sci 24(3):212–218, In Chinese

Index

© Springer Science+Business Media New York 2016

A. McElhatton, M.M. El Idrissi (eds.), *Modernization of Traditional Food
Processes and Products*, Integrating Food Science and Engineering Knowledge
Into the Food Chain 11, DOI 10.1007/978-1-4899-7671-0

Printed in the United States
By Bookmasters